Down and Dirty!
A Kit of Activities in Environmental Science

Marshall L. McCall
and
Carl E. Wolfe

York University
Toronto, Canada

KENDALL/HUNT PUBLISHING COMPANY
4050 Westmark Drive   Dubuque, Iowa 52002

Cover image © Mario Lopes, 2008
Under License from Shutterstock, Inc.

Copyright © 2005, 2008 by Marshall L. McCall and Carl E. Wolfe

ISBN 978-0-7575-5644-9

Kendall/Hunt Publishing Company has the exclusive rights to reproduce this work,
to prepare derivative works from this work, to publicly distribute this work,
to publicly perform this work and to publicly display this work.

All rights reserved. No part of this publication may be reproduced,
stored in a retrieval system, or transmitted, in any form or by any
means, electronic, mechanical, photocopying, recording, or otherwise,
without the prior written permission of the copyright owner.

Printed in the United States of America
10 9 8 7 6 5 4 3

Table of Contents

Preface v

Read This First! 1

Materials 11

Activity 1: Biodiversity
    Counting Large Populations 13

Activity 2: Overpopulation
    Exponential Growth 27

Activity 3: Global Warming
    The Enhanced Greenhouse Effect 37

Activity 4: Global Refrigeration
    The Albedo 43

Activity 5: Ozone Depletion
    Ultraviolet Light 49

Activity 6: Energy
    Solar Power 55

Activity 7: Air Pollution
    Acid Deposition 67

Activity 8: Intensive Agriculture
    Soil Quality 73

Activity 9: Genetically Modified Organisms
    Genetics 81

Activity 10: Radiation
    Biological Effects of Radiation Exposure 89

Preface

One of the most challenging aspects of teaching a science course to a large class of non-science majors is finding ways to expose students to hands-on activities... so challenging, in fact, that it appears it is often not done. Indeed, many textbook publishers appear to view "hands-on" to mean use of the World-Wide Web.

Annually, we teach a course in environmental science to up to 500 non-science students. Often, because of previous miserable experiences in the school system, many of the students are leery about science, if not downright afraid of it, and only take the course to meet a credit requirement. Our objective has been to make science understandable, meaningful, and fun, and at the very least, to have students leave with some appreciation of science and the way scientists think.

To achieve our goals, we felt that it was important to have students actually do experiments. However, ours is not a lab course and laboratory space for such large classes is unavailable. Instead, we run tutorials at which selected topics are elaborated upon and students are given a take-home assignment. It is through these tutorials that students can gain real experience with the ways of science.

Initially, we sought out inexpensive hands-on activities by surveying the Web and speaking to many publishers. Soon, we came to the surprising realization that no affordable package of exercises *and* materials to do them existed. Thus, we began to devise our own activities – constrained, of course, by a minimal budget. It was found that some practical experience with experimentation and the scientific method could be conveyed via take-home assignments requiring inexpensive supplies like thermometers or pH paper. However, besides the logistical problem of distribution and recovery, we found we were losing 20% of our supplies annually.

Given our frustrations, we became very excited when Kendall-Hunt suggested putting together an activity kit for our students. We were at once able to revisit ideas that had previously been discounted as either too costly or too cumbersome. The end result of the process is this book and the accompanying kit, which were first published in 2005. This version, the second edition, fills gaps in the original by including two new activities on energy and radiation.

Although intended for use in a very specific environmental science course, we believe that this book also has value outside of that context. Indeed it is, to the best of our knowledge, the first attempt to package hands-on at-home activities suitable for the practical component of a large class in environmental science.

Each activity is centered around a major environmental concern, but focuses on a particular scientific concept or method relevant to the main issue. After providing background (which we also go over in lectures, ultimately), each activity is laid out in such a way that the student is largely left to his/her own devices to accomplish the stated goals. We have deliberately chosen this approach to foster a spirit of creativity in experimental design, and thereby encourage curiosity and learning. Nevertheless, numerous hints and tips are provided with each activity, and where a particular technique is the object of study or simply crucial to the stated goals, more detailed directions are given.

Normally, we give students two weeks to carry out each activity. However, the activity associated with *Radiation* requires students to grow radishes to maturity, which takes about four weeks. Thus, it is recommended that students be alotted six weeks in total to complete the activity. This requires that the project be assigned near the beginning of a semester. For students to complete the activity associated with *Energy*, it is necessary that the Sun be unobscured by clouds. In poorer climates, the instructor may have to give students more than two weeks to complete the activity, or at least be prepared to give students an extension if the weather fails to cooperate.

The activities are laid out in the order in which we teach the subject matter, although enough background is provided in each that lectures need not be synchronized with the activities (which is almost impossible for a large class with multiple tutorial sections). After delving at length into the scientific method, we focus on ecological issues, then proceed to concentrate on the science underlying environmental phenomena. We follow with applications. Although it is possible to change the ordering, it is important to realize that certain pairs of activities are linked. *Global Warming* should precede *Global Refrigeration*, because the latter relies heavily upon the information about temperature, steady states, and the greenhouse effect given in the former. Also, it is advisable to schedule at least one other activity less dependent on experimental design between these two, such as *Biodiversity*. This will ensure that there is enough time to give students graded feedback on the methodology they adopt for Global Warming, which has a direct bearing on how they perform the activity on Global refrigeration. It is important that *Ozone Depletion* precede *Energy*, because the fundamentals of electromagnetic radiation laid down in the former are needed to address the subject of solar power covered in the latter. *Air Pollution* should precede *Intensive Agriculture*, because the latter relies upon knowledge about pH and its measurement, all of which is presented in the former. *Biodiversity* should precede *Genetically Modified Organisms*, because the microscopic picture of life given in the latter draws from knowledge about the macroscopic picture described in the former. *Radiation* ought to follow *Genetically Modified Organisms*, because students are expected to be familiar with concepts underlying the molecular basis of life which are provided as background to the latter.

We hope that the kit fosters a spirit of exploration in all who use it.

Marshall McCall
Carl Wolfe

York University
Toronto, Canada
May, 2008.

Acknowledgments

We thank Neil T. McCall for his idea to introduce an activity on the enhanced greenhouse effect and for his help with growing radishes, and Susan H. C. P. McCall of Stellar Optics International Corporation (SORIC) for her advice on black and white surfaces and materials and for her help in testing albedos. Paul Mortfield made us aware of UV-sensitive beads. We are also grateful to Priscilla W. But for helpful comments and for lending a teacher's eye to parts of the manuscript. Over the years, we have also benefited from interactions with many talented students and teaching assistants who have conveyed their views about our approaches to teaching environmental science.

Read This First!

What Is the Environment?

The environment is what surrounds you. It extends from your skin all the way to the edge of the universe. However, the part which is relevant to your well-being or the well-being of others depends upon your perspective. It could be your pillow. It could be the water you drink. It could be your garden. It could be the air you breathe. It could even be the solar system.

The environment sustains life. Unfortunately, parts of it are under stress, and its ability to sustain life is threatened. The better you understand the environment, the better you will be able to defend it and your well-being. The path to understanding is science.

What Is Science?

In its purest form, science is the pursuit of knowledge about the natural world. With that knowledge follows applications, the development of which also may be regarded as science. As with any human endeavour, there are good ways and bad ways to do science. The bad ways can lead to incorrect or exaggerated assertions, which can be misleading at best and harmful at worst. Thus, it is worth your while to learn about the good ways.

The good ways are guided by a kind of **inductive reasoning** known as the **scientific method**. Inductive reasoning is a bottom-up kind of thinking where a general conclusion is derived from limited observations. Fundamentally, the scientific method works like this:

Observe
Make an observation about the world.

Question
Ask why the world works this way.

Hypothesize
Suggest an explanation, i.e., a **hypothesis**.

Experiment
Carry out a *useful* experiment designed to test the hypothesis. The definition of *useful* is given below.

Conclude
From the result of the experiment, conclude whether or not the hypothesis is supported.

Disseminate
Tell the world about your findings, normally in a peer-reviewed scientific publication.

If a hypothesis is shot down by the result of an experiment, it means "back to the drawing board", that is, time to devise a new hypothesis. On the other hand, if a hypothesis is supported by the experiment, it gains greater credibility. If a hypothesis is repeatedly supported by many different experiments, it gains the status of a **theory**. If it withstands tests over a very long period of time, it may gain the status of a **law**. A law is the strongest scientific statement possible, and represents the closest thing to a "fact." However, it is not really a fact, because experimental testing is never complete. For example, the law of gravity would have to be revised the day that a human is observed to fall up rather than down.

In some ways, the scientific method amounts to common sense. Without knowing it, you may already be practicing the scientific method in everyday life. Here is an example. Suppose you are driving along a road with your friend. Up ahead, you **observe** a dog lying on its side just off the edge of the road. You **question**, "What is wrong with the dog?" You then **hypothesize** that the dog is dead. Being a good scientist, you pull over to the side of the road to conduct an **experiment**. You get out of the car and cautiously approach the dog. As you get close, you notice that the dog is breathing. You have disproved your hypothesis. You **conclude** that the dog is alive. You **disseminate** this information to your friend by yelling, "The dog is alive!" You pose a new **question**: "Why is a live dog lying at the side of the road?" You then **hypothesize** that the dog is hurt. You conduct an **experiment**: you look for signs of damage. You find none. You then conduct another **experiment**: you poke the dog gently with a stick. The dog jumps up and licks you. You **conclude** that the dog is all right, disproving your hypothesis. You **disseminate** the information to your friend by yelling "The dog looks OK!" You ask another **question**: "Why was a healthy dog lying at the side of the road?" You come up with another **hypothesis**: "The dog was sleeping." However, you decide to end your research at this point by telling your friend that the matter ought to be further investigated by others, such as a veterinarian.

The scientific method is the overwhelmingly dominant approach to doing science today. However, there is another way of doing science which is legitimate under some circumstances. Scientists working at the frontier of human understanding have little or no knowledge about what they are investigating. The focus must be on observing. Thus, experiments are carried out to provide information, not to test a hypothesis. A fact-finding approach like this is called **exploratory science**. The scientific method kicks in once the knowledge gap has been narrowed. Exploratory science was commonly practiced in the past. For example, Michael Faraday initially followed this approach when he started his work on electricity and magnetism. Today, science is results driven, and exploratory science is frowned upon. If a scientist were to apply for a grant to carry out exploratory research into a phenomenon, he/she likely would be accused of going on a "fishing expedition" and be refused the money. In your case, though, an exploratory approach to science sometimes may be of considerable benefit, since so many issues are at the frontier of your knowledge. Certain activities in this book will open up this path to you.

What Is Environmental Science?

Environmental science, as the name suggests, is an approach to studying the environment founded upon the scientific method. It is **interdisciplinary**, in that it draws upon *many* fields of human knowledge (e.g., astronomy, biology, chemistry, physics, history, philosophy, economics, and politics). Also, it is **observational**, **theoretical**, and **applied**. Observational scientists conduct experiments to test hypotheses about how the environment works. Theoretical scientists amalgamate the results of many experiments to develop strong statements (theories) about the way the environment works. Having established these statements, applied scientists try to use the them to solve real-world problems.

As an example, in 1985 observers found that the amount of ozone in the atmosphere was declining, in turn increasing our exposure to harmful ultraviolet light from the Sun. Theoreticians showed that propellants in spray cans were extremely effective in destroying ozone molecules. Then, applied scientists devised alternative propellants less harmful to the environment.

Experimentation

Bad versus Good

How an experiment is designed determines whether or not it will provide a useful test of a hypothesis. To understand what a good experiment is, consider first a bad one. Suppose you were to hypothesize that eating hamburgers makes hair turn grey. To test your hypothesis you feed your parent a hamburger. You notice that your parent's hair starts turning grey two years later. You conclude that eating hamburgers does indeed cause hair to turn grey, and proceed to send a letter to your local newspaper to inform the world of this fact.

What is wrong with this picture? Lots. To deliver a useful test of a hypothesis, an experiment must be *controlled*, *statistically significant*, and *repeatable*.

The Controlled Experiment

The outcome of any action may be determined by many factors, each of which is known as a **variable**. A controlled experiment is one where the outcome is determined to the best of one's knowledge by *only one variable*. This is accomplished by performing the action on a **test group** of subjects or things and *not* performing it on a **control group** ("control" for short) of subjects or things which is identical as far as is possible to the test group. Then, any response of the test group which is different from that of the control group must be due to the action. To test the greying hypothesis, you should compare the response of the parent fed a hamburger to that of another parent *not* fed a hamburger, the second parent having the same age, sex, health, eating habits, living environment, working environment, etc., as the first.

The Statistically Significant Experiment

A statistically significant experiment is one where there are enough subjects in the test and control groups to ensure that any difference in the response of the groups to the action on one of them is unlikely to be a fluke. To test the greying hypothesis, you should feed a hamburger to each of 1000 parents (the test group) and not feed a hamburger to each of another 1000 parents with similar characteristics (the control group), and monitor the responses of the two groups. A statistically significant verification of the hypothesis would arise if 500 parents in the test group turned grey but only 100 in the control group did. On the other hand, you might think twice if 101 parents in the test group turned grey compared with 100 in the control group, as the difference might purely be an accident.

Statistical significance may also be gained by conducting an experiment more than once. Inherent in any experiment are uncertainties associated with the group selection or the process of measurement. Repetition can help to identify how sensitive your results are to these uncertainties, and thereby give you a quantitative measure of the confidence with which you can draw conclusions. For example, to test the greying hypothesis, you might consider repeating your experiment with two new groups of people.

Keep in mind that what comes out of an experiment is only as good as what goes in; if your experimental design is flawed, the outcome will be flawed every time the experiment is repeated. For example, say you have a ruler which is marked as being 30 cm long. Suppose you measure the length of a piece of wood three times, and find good agreement among the measurements. This will boost your confidence that you haven't messed up the measurement. However, if your ruler is actually 40 cm long (but marked as 30), then your answer would be wrong because the measurements were **systematically** wrong. Since you needed a precise measurement of length, you should have considered measuring the length in different ways, say by using different kinds of rulers, or even a laser range-finder.

The Repeatable Experiment

A good scientist trusts no one. After coming up with a conclusion, especially a controversial one, or after developing a technique for doing something, you can expect that there will be others who will want to verify it. To do so, it may be necessary to conduct the experiment again in the same way. A repeatable experiment is one whose construction is so carefully described that it can be repeated by others. Naturally, it is expected that such an experiment will yield a similar outcome. Repeatability is essential to determining reliability, as it helps to overcome bias, mistakes, and uncertainties, especially of a systematic nature. In the case of the greying hypothesis, it would be desirable for an independent scientist to repeat your study, but with two different groups, perhaps larger than yours, in a different location.

Record-Keeping

In 1922, Charles Best, a Canadian undergraduate student assisting insulin co-discoverer Dr. Frederick Banting, led a team which developed a potent, pure extract of insulin suitable for treating diabetes in humans. He was not the first to do so, yet he received the credit. Why? A competitor of Best, Dr. James Collip, managed to create a pure extract first, which in fact was used in the first successful treatment of human diabetes. However, he failed to keep adequate notes of how he did it. He was unable to repeat what he had done, and of course couldn't tell anyone else what to do, either.

Perhaps the most important aspect of experimentation is the keeping of records. Whenever an experiment is done, it is imperative to record who did it, when it was done, and where it was done, as well as the details of how the experiment was conducted and any measurements which were made. This information is called a **log**. *Record-keeping is not something that should be done after the fact.* It should be done as the experiment is conducted. To avoid accidental loss of important information, notes should be written with ink, not pencil. Completeness, not neatness, should be the priority. An an example, a log of research on the dog lying at the side of the road is given in Figure 1.

Uncertainty

Uncertainty is an intrinsic part of the measurement process. For example, even if you had a perfect ruler, measurements of length still could not be made perfectly, because there is a degree of randomness associated with the positioning of the ruler and the reading of the ruler. For example, limitations in your vision, unsteady hands, and ill-defined edges all may compromise the measuring process. In any experiment, there are many factors over which you have little control, and these can lead to answers which deviate from the truth in random ways. Such random deviations are referred to as **random errors**. They are different from the **systematic errors** described previously in that deviations may be high or low with respect to the true value. Thus, repeated measurements only afflicted by random errors will be distributed around the correct answer. In contrast, systematic errors cause deviations to one side or the other of the correct answer.

Suppose you wanted to measure the size of your pet turtle. To gain statistical significance, you would logically make several measurements and compute the average, namely the sum of all measurements divided by the number of measurements. Say you make four measurements with your ruler, and find lengths of 5, 6, 6, and 8 cm (the little fellow was squirming). The average would be $(5 + 6 + 6 + 8)/4 = 6.25$ cm.

With the average, it is important to convey a sense of the difficulty you had in making your measurements, namely an estimate of the uncertainty due to random errors. One way is to compute half the difference between the maximum and minimum values. For the size of your turtle, the uncertainty would be $(8 - 5)/2 = 1.5$ cm.

The average *plus or minus* the uncertainty defines a *range* within which the true answer is likely to lie. You would express your answer for the size of your turtle as 6.25 ± 1.5 cm. What this says is that, given the difficulty in making the measurements, your best guess for the size of your tur-

Read This First!

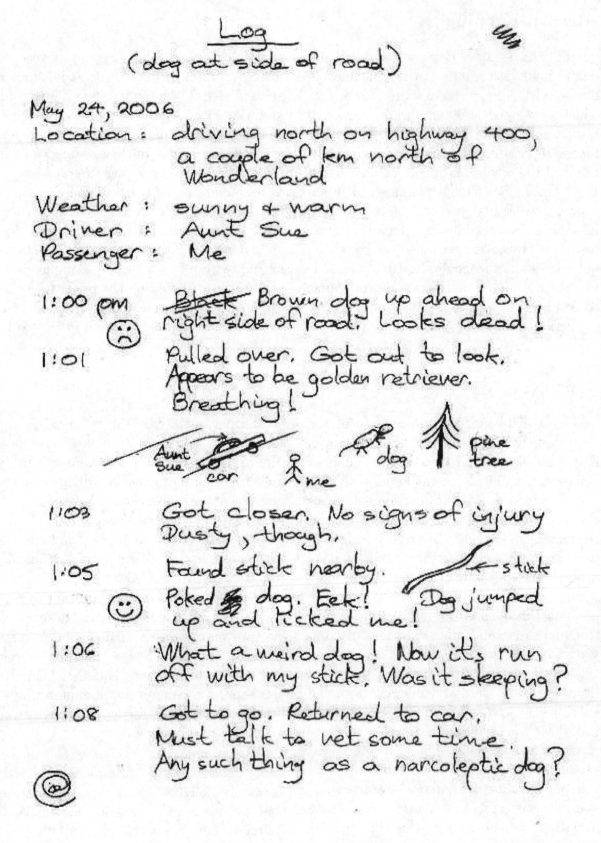

Figure 1: Example of a research log.

tle is 6.25 cm, but that it is reasonable to expect that your turtle could be as small as $6.25 - 1.5 = 4.75$ cm or as large as $6.25 + 1.5 = 7.75$ cm. It is important to understand that scientific uncertainty doesn't imply a mistake, but rather is an honest appraisal of the reliability of the final answer.

Graphing

A vital part of the scientific process is the graph. A **graph** is a visual representation of measurements which is used to identify and analyse relationships.

For example, suppose that you measure the size of your pet turtle every month for a period of twelve months. Each measurement would be recorded as a pair of numbers, namely time and size. The easiest way to summarize the growth of your turtle is with a graph showing it's size as a function of time. Such a graph, with key parts highlighted, is shown in Figure 2. Since the size of the turtle *depends* on time, it is conventional to plot the size along the vertical axis and the time along the horizontal axis.

Each **point** on the graph refers to a specific measurement of time and size. The time is given by the projection from the point down to the horizontal axis. The size is given by the projection across to the vertical axis.

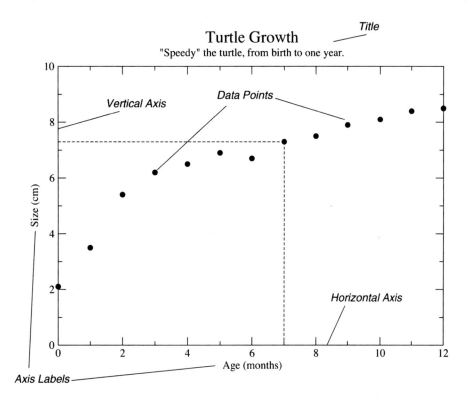

Figure 2: Example of a graph and definitions of its major parts. The points (black circles) show the outcomes of individual measurements of the size of your pet turtle. The dashed lines show how one of the points is positioned. The projections to the horizontal and vertical axes give, respectively, the time and size associated with the point.

Read This First!

Dissemination

After going to all of the trouble of performing an experiment, it is only appropriate that you disseminate your results so that others may learn from your experience. Usually, that means writing up a summary. Any write-up should contain:

- A discussion of context, namely background information as to why the work is important in the grand scheme of science.

- A statement of the hypothesis being tested.

- A description of the experiment, especially the setup, with diagrams where useful.

- A presentation of your measurements (e.g., tables, graphs, etc.).

- A description of how your measurements were made and the associated uncertainties.

- Conclusions about the viability of the hypothesis and implications for the world.

Where should you send your write-up? Normally, it should be submitted to a refereed scientific journal, whereupon it will be passed on to one or more scientists in your field to review. If your work is deemed good, it will be published; if not, it will be rejected. In your case, where you send your write-up may depend upon what somebody else, like your instructor, tells you. However, the last place you should send it to is the media. Many journalists do not have scientific training and are unable to judge the credibility of a scientific report. Worst of all, their attention is selective. By not presenting the work in its entirety, there is a danger that the conclusions will be distorted. Plus, if you have made a mistake, the whole world will end up knowing. It is best to give the scientific community a chance to judge your work first before revealing it to the world.

Is There a Right Answer?

The outcome of an experiment leads to an answer about the viability of some hypothesis. Sometimes the answer may be in conflict with current wisdom, in which case it will be controversial. However, that doesn't necessarily make it "wrong." The credibility of the answer, that is, rightness, will be judged on the basis of how well the experiment was designed and conducted. For example, if you performed a controlled, statistically significant, and repeatable experiment which demonstrated that eating hamburgers turns hair grey, then the result would have to be taken seriously. Whether or not this is an absolute truth, only Nature knows for sure.

What You Have before You

This book presents a selection of activities designed to give you hands-on experience with environmental science. In many instances, you will need to design an experiment to test a hypothesis. We don't tell you how to to design the experiment, partly because there is more than one way to make a useful experiment, and partly because you will learn better if you do it yourself. However, having learned above about the scientific method and the process of experimentation, you are well-armed to conduct useful experiments. Just think: controlled, statistically significant, repeatable.

Each activity is introduced with some background material describing the science behind it. Usually, a hypothesis is then proposed, and you are asked to design and execute an experiment to test it. To help you conduct the experiment, recommended materials are listed and tips on experimental procedure are offered. A log sheet is supplied for record-keeping. So, with the aid of your kit, get down and dirty and do some science!

Materials

What Is Included in Your Kit

- 30 grams of rice in a clear plastic bag
- 1 opaque plastic bag
- 1 ruler
- 1 sheet of grid paper (included as a tear-out in this book)
- 5 sheets of graph paper (included as tear-outs in this book)
- 2 thermometers
- 60 grams of salt in a clear plastic bag
- 20 UV-sensitive (colour-changing) beads
- 1 beading string
- 2 samples of sunscreen
- 1 solar cell
- 1 ammeter
- 2 wires with alligator clips on the ends
- 10 pH test strips
- 4 empty clear plastic bags
- 4 coffee filters
- 4 rubber bands
- 1 plastic spoon
- 1 plastic pipette
- 3 clear plastic test tubes with caps
- 5 nitrate test tablets
- 5 phosphate test tablets
- 10 or more radish seeds which *have not* been exposed to radiation from cobalt-60 (distinguished by being in a *white* plastic bag)
- 10 or more radish seeds which *have* been exposed to radiation from cobalt-60 (distinguished by being in a *coloured* plastic bag)

What You Will Need to Find

- 1 dark felt-tipped marking pen (for colouring rice grains)
- At least 2 identical glass or clear plastic containers with lids (e.g., spaghetti sauce jars)
- A ground cover (for the inside of a container) of your choosing
- UV-blocking sunglasses (if available)
- 1 large container for collecting rain water or snow, such as a bucket
- 2 samples of dry soil taken from different locations (from a garden and a potted plant, for example)
- 1 measuring cup (optionally)
- At least 2 containers with holes in the bottom along with 2 pedestals for growing plants indoors
- Soil suitable for growing plants indoors.
- Plastic wrap

Activity 1
Biodiversity
Counting Large Populations

What Is Biodiversity?

Simply put, **biodiversity** refers to the richness of life. Richness is conveyed by both the number of different life forms as well as the range of characteristics exhibited by them. Different life forms are called species. Two organisms are said to be members of the same **species** if they can reproduce freely in the wild. The entire collection of organisms making up a species is called a **population**.

Biodiversity takes two forms:

1. **Species diversity** refers to the variety of species.

2. **Genetic diversity** refers to the variety of traits within a species.

A **trait** is a physical characteristic that is inheritable through reproduction. At the molecular level, traits are determined by genes. A **gene**, which is a segment of a large molecule (DNA), provides instructions for the manufacture of a chemical underlying a trait. Genes will be the subject of a later activity.

Genetic diversity is the result of natural random changes in genes called **mutations**, which lead to new traits. According to Charles Darwin's Theory of Evolution, organisms typically produce more offspring than can survive in a given habitat. The result is competition between organisms for food, space, mates, and so on, all of which are in limited supply. Those organisms whose particular traits promote the survival of offspring are said to be well **adapted** to their environment, and flourish. Organisms with traits which impair the survival of offspring are less well adapted, and wither. For example, green frogs are better adapted to their environment than fluorescent pink frogs, because predators have a harder time finding them. Those organisms best adapted to their environment are said to be the **fittest**. The fittest survive!

It follows that traits that are good adaptations are passed on to future generations. Traits that are not become more rare because the carriers are disadvantaged (witness fluorescent pink frogs). This process is called **natural selection**.

When all members of a species have died, the species is said to have become **extinct**, and there is a loss of biodiversity. Extinction, it should be noted, is a natural part of the evolutionary process. It is estimated that, under normal conditions, one species out of every thousand becomes extinct every thousand years. However, human intervention in the environment is accelerating the rate of extinction, particularly as unique habitats, such as rainforests, are encroached upon.

Why Is Biodiversity Important?

Biodiversity has aesthetic value by mere virtue of its existence. It also has practical value. Genetic diversity is important for the long-term health of a species. For example, it helps to protect the very existence of the species from being threatened by a single disease. Even if only 1% of the population carries a gene describing how to defend the organism against a disease, this may help the species to survive an outbreak. Future generations are, in this way, shielded through natural selection.

If the preceding seems abstract and disconnected from human reality, then consider that around the world today only a small number of varieties of rice, wheat, and corn are grown for food. In the past, several hundred varieties, each native to different regions of the world, were grown. Clearly there has been a loss of diversity in food crops. The result is that today billions of people depend on just a few genetic varieties for their staple foods. Should a serious disease attack one of these varieties, then a huge part of the world's food supply would potentially be in jeopardy. Fortunately, there are over a thousand seed banks (or gene banks) located around the world that preserve seeds of local crop varieties in case of just such an emergency.

Biodiversity should be regarded as a precious resource. The genes that determine traits are the outcome of millions of years of evolution. We can't yet custom-make genes ourselves. However, through genetic engineering, we can make use of existing genes to solve real-world problems. For example, scientists have inserted a gene from spinach into the DNA of a pig to develop a leaner and healthier pig with less saturated fat.

Now, suppose that the cure for the common cold currently resides in a unique trait of a particular variety of beetle somewhere in the Amazon rain forest. Further suppose that, because of new road construction, the beetle's natural habitat has been steadily eroded to the point that there are now only two such beetles in existence anywhere, and both are male. The species is now on the verge of extinction and when the last individual dies the cure for the common cold will vanish with it.

The preceding is an extreme example, but it illustrates the reasoning behind the drive to preserve biodiversity. Useful traits may be out there awaiting discovery; but if the species that carry them die out, then those traits will be gone forever.

Monitoring Biodiversity

To monitor the health of a species, ecologists often face the daunting task of counting large numbers of individuals. For example, look at an ant hill. How many ants make it their home? Similarly, how many blades of grass are growing in a lawn? Of course, both of these questions can be answered by laboriously counting each and every ant and each and every blade of grass. But doing so is very time-consuming. Moreover, the high likelihood of mistakes means that these questions can never be answered *exactly*. However, ecologists use a number of methods to obtain *estimates* of the size of populations that require far less time and far less work.

Immobile Individuals

If the individuals being counted are fixed in place within a defined region of study (e.g., flowers in a field), then the total number of individuals can be estimated by considering a subset of the population, called a **sample**, drawn from a small part of the region. The number of individuals in that part divided by its surface area defines a **surface density** which, when multiplied by the total surface area of the region, yields an estimate for the size of the population.

To see how this works, let us return to the lawn. Suppose that you identify a 5 cm by 5 cm square of lawn and actually count the number of blades of grass in that square. Let's say you count 75 blades. Then in that specific location, there are 75 blades of grass in a 5 cm × 5 cm = 25 square centimetres (cm^2) area, or 3 blades per cm^2. If the lawn measures six metres by ten metres, then the area is 600 cm × 1000 cm = 600,000 cm^2. You can immediately deduce that there must be approximately 600,000 cm^2 × 3 blades per cm^2 = 1,800,000, or 1.8 million blades of grass in the lawn.

In general, to count the number of immobile individuals (N) of a particular species inside a defined region of total surface area A, we count the number of individuals (n) in a smaller area (a). We expect

$$\frac{n}{a} = \frac{N}{A}.$$

Multiplying both sides by A, the total number of individuals in A is given by

$$N = A \times \left(\frac{n}{a}\right).$$

This means that only comparatively few individuals have to be counted to get an estimate for the total size of the population.

Implicit in the method just described is the assumption that the density of individuals measured in a small part of the region of study is a good reflection of the overall density. Just in case it isn't, it is preferable to analyse more than one part and to take an average. For example, it wouldn't be sensible to estimate the number of blades of grass in a lawn by counting blades in a bare patch.

Mobile Individuals

Now suppose that you decide to count all of the ants in an ant hill. You will quickly face a very big challenge because ants move, and they probably do so more quickly than you can count them. Fortunately there is an easier way that actually is aided by this movement. It is called **mark-recapture** estimation.

Mark-recapture (or Petersen) estimation is founded upon simple probability. For example, say you have a total of ten socks in a drawer, eight of which are black and the remaining two of which are green. If you blindly reach into the drawer and take out one sock, then eight times out of ten that sock will be black and two times out of ten it will be green. In other words, for any drawing of one sock there is an 80% chance that it will be black and a 20% chance that it will be green.

If you conduct an experiment in which you draw a sock 30 times (each time putting the sock back in the drawer and mixing all of the socks up) then you *expect* to draw a green sock six times (20% of 30 times). Of course, in any given experiment you may draw the green sock more or less often than this, but as the number of draws increases and as the experiment is repeated many times, the *average* number of green socks drawn gets closer and closer to six.

Getting back to the ant hill example, suppose for a moment that we know there are exactly 500 ants in the ant hill. Suppose also that we capture 67 ants and label (or tag) each of them with a small dab of green paint before releasing them back into the ant hill. Because the ants move around quickly, we can safely assume that after a short time (say a day) the tagged ants have thoroughly (randomly) mixed with the untagged ones. However, we do know that there are 67 "tagged" individuals among the 500 ants in the ant hill. The odds of capturing a tagged ant are, therefore, 67 in 500, or about 13%. If we capture 45 ants, then the expected number of tagged (or recaptured) ants among those captured is, on average, 13% of 45, or 6.

The prescription that emerges is that the expected proportion of tagged ants in any given sample of the population is equal to the ratio of the total number of tags in the population (T) to the size of the population (N). In other words, for a sample of size n containing t tags,

$$\frac{t}{n} = \frac{T}{N}$$

This expression immediately suggests a solution to our problem, namely the estimation of N. Re-arranging,

$$N = T \times \left(\frac{n}{t}\right)$$

So, if we have T tags in a population we can estimate the size of the population by drawing a sample from the population and counting how many in this sample are tagged.

What You Should Do

It is claimed above that the number of individuals in a population can be estimated using two different methods. How can the reliability of either be tested? One way would be to apply them to a situation where the number of individuals is known exactly. But that wouldn't be very realistic. Instead, we propose the following hypothesis:

> *Where applicable, the surface density and mark-recapture methods yield answers that are consistent with one another within uncertainty.*

To give you practice with estimating the size of a population we supply you with a bag of rice. The rice grains take the place of living organisms, be they ants or blades of grass. You can test the hypothesis by using both the surface density method and the mark-recapture method to estimate the total number of rice grains in the bag. Note that at no time will you be told how many grains are actually in the bag. Even we, the authors of this book, don't know the exact answer to that question.

For the surface density method, spread the rice over the coarse grid paper provided with this activity. You can consider each square to represent one "unit" of area. The exact size of each square is irrelevant to the exercise. To estimate the total surface area, count the number of squares (units) covered by all of the rice, including any partial (half, quarter) squares. Count the number of rice grains in one randomly-selected square. Then, using your previous estimate of the total surface area, work out the total number of rice grains. Repeat this calculation as many times as you see fit using counts for different squares.

For the mark-recapture method, tag some of the rice grains with a dark felt-tipped marking pen in such a way that they can be easily spotted. Keep an accurate count of the number of tagged grains as this figure is critical to your estimates. You should aim to tag about 100 grains but the exact figure is up to you. Then, put all of the rice in the opaque plastic bag. Close and shake the bag vigorously. Without looking in the bag, use the spoon provided in the kit to draw a sample of rice. By feel alone, aim for about half a spoonful. After counting the number of rice grains in the sample as well as the number of tagged grains, return the sample to the bag. Shake the bag again. Repeat the sampling and counting process as many times as you see fit. For both methods, quantify the uncertainty in the estimate of the total number of rice grains.

What You Will Need

From Kit

- Rice
- Opaque plastic bag
- Plastic spoon
- Ruler
- 1 sheet of grid paper (tear-out included with this activity)
- 2 sheets of graph paper (tear-outs included with this activity).

From Other Sources

- Dark felt-tipped marking pen

Things You Should Think About

- For the surface area method, keep in mind that one of the assumptions made is that the average density is about the same everywhere.

- Calculating the total surface area will be easier if you try to arrange the rice grains over a rectangular area of the grid paper. Consider using a ruler to spread and "herd" them.

- For the mark-recapture method, you may want to explore the effect on your estimates of increasing the number of tagged rice grains.

What You Should Disseminate

Page 1
Describe your experiments and your findings.

Pages 2 and 3
For each experiment, show your estimates for the total number of rice grains visually by sketching a graph on the graph paper provided with this activity. Plot the trial number on the horizontal axis and the estimated number of grains on the vertical. Also plot the average as a dashed horizontal line.

Page 4
Your original log sheet. This should include a table showing all of the measurements for each of the methods.

Page 1 should include:

- A statement of the hypothesis being tested.

- A description of each experiment you performed, including the number of tags used and the number of trials.

- Your conclusions concerning the hypothesis and your best estimate of the total number of rice grains in your kit along with the uncertainty.

Activity 1   Biodiversity

Log Sheet: Counting Large Populations

Activity 1   Biodiversity

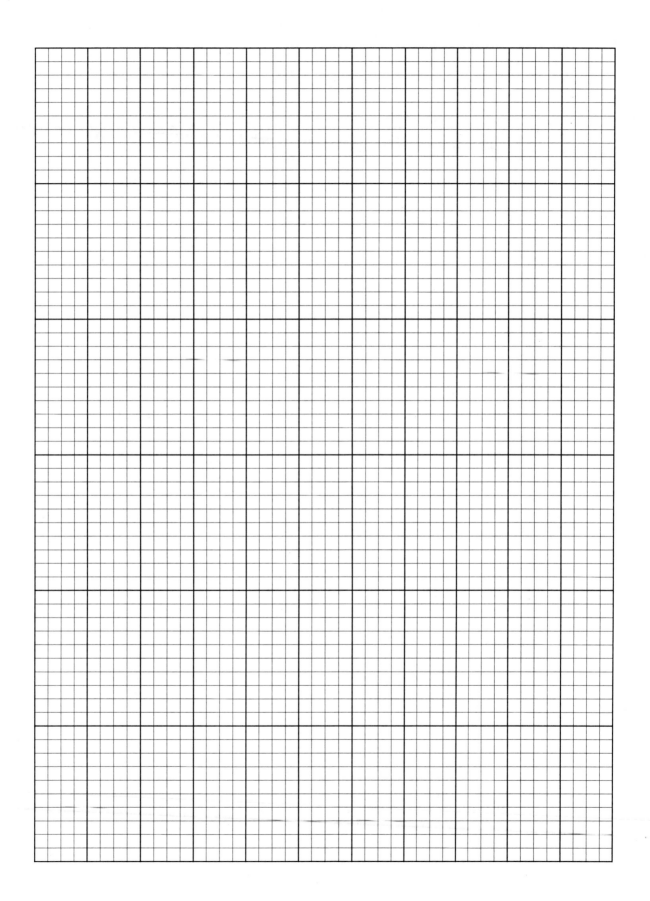

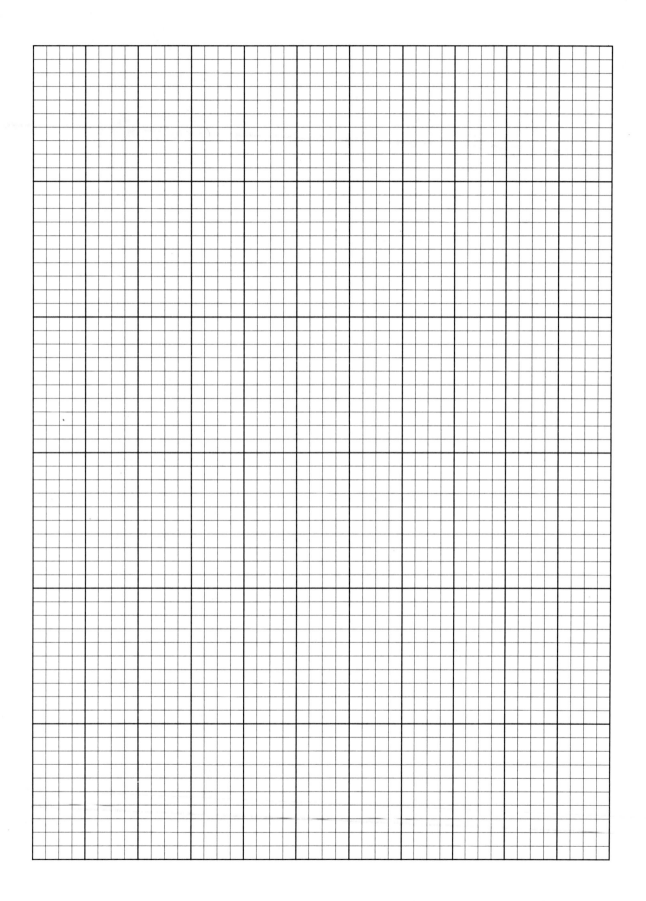

Activity 1 Biodiversity

Activity 2
Overpopulation
Exponential Growth

What Is Overpopulation?

Individuals of a species living within a defined ecosystem draw their food, water, and any other resources needed to sustain themselves primarily from that ecosystem. If the demand for resources is small compared to the total supply of resources, then in the absence of predators that species will flourish. The number of individuals in the population, which is itself usually referred to as the **population**, will increase.

However, as the population increases, there comes a time when the supply of resources available to each individual starts to decline. This happens because a limited resource must now be shared among an ever increasing number of individuals. In other words, there are now more individuals than the available resources can sustain. The ecosystem experiences an **overpopulation** of the species in question. In nature, the end result of overpopulation is usually the starvation and death of large numbers of individuals until a balance is restored between the population and the food supply.

What Is Exponential Growth?

In a factory or on an assembly line, the number of units of a product made in any interval of time, say an hour, is constant. So, the total number of units produced grows in proportion to the amount of time passed. For example, if a bakery makes 30 cakes per hour, then the total number of cakes made by the bakery grows during one day as shown in the second column of Table 1 below. This kind of growth is called **linear growth**.

In contrast, the population of living organisms exhibits what is called **exponential growth**. This kind of growth is best illustrated by single cells such as bacteria, which reproduce by dividing into two after a certain time interval. Starting from one cell, there will be two after one interval of time. After a second interval of time has passed, each of the two cells divides into two once again, leaving a total of four cells. And so on. Assuming that the time interval between cell divisions is one hour, then the total number of bacteria should grow as shown in the third column of Table 1 (Fast Bacteria).

In general, exponential growth involves a particular **growth rate**, defined as the fractional increase that takes place in one time interval. For the bacteria example the growth rate is 100% in one hour, because after one hour there has been a gain in bacteria *equal* to the number of bacteria at the start of the hour. For larger organisms growth rates are much smaller. The human population, for example, is growing at a rate of roughly 1.25% per year. That means that each year we add 1.25% of the current population to the total. If P_0 is the population in some year, then the population P_1 a year later is

$$\begin{aligned} P_1 &= P_0 + P_0 \times (0.0125) \\ &= P_0 \times (1 + 0.0125) \\ &= P_0 \times 1.0125 \end{aligned}$$

Table 1: Cakes, slow-dividing bacteria, and fast-dividing bacteria. It is assumed that 30 cakes are made every hour, fast bacteria grow in number at a rate of 100% every hour, and slow bacteria grow in number at a rate of 50% every hour.

Time	Cakes	Fast Bacteria	Slow Bacteria	Time	Cakes	Fast Bacteria	Slow Bacteria
1 PM	0	1	10	6 PM	150	32	76
2 PM	30	2	15	7 PM	180	64	114
3 PM	60	4	22	8 PM	210	128	171
4 PM	90	8	33	9 PM	240	256	256
5 PM	120	16	50	10 PM	270	512	384

After two years of growth at the same rate the total population is given by

$$P_2 = P_1 + P_1 \times (0.0125)$$
$$= P_1 \times (1 + 0.0125)$$
$$= P_0 \times (1 + 0.0125) \times (1 + 0.0125)$$
$$= P_0 \times 1.0251$$

So, in the general case of a constant growth rate of r percent in some time interval (typically a year), the total population P_N after N intervals is

$$P_N = P_0 \times \left(1 + \frac{r}{100}\right)^N$$

To demonstrate how sensitive population growth is to the growth rate, the population of a different species of bacterium (Slow Bacteria) with a lower growth rate of $r = 50\%$ per hour (instead of 100%) is tracked in the fourth column of Table 1. In this case, a starting population of ten individuals is assumed. Even though there are ten times more slow-dividing bacteria at the start, the population of fast-dividing bacteria ends up predominating after only 8 hours.

Why Does Exponential Growth Matter?

Environmental problems today stem from two factors:

1. The impact of each human being on the environment.
2. The number of humans on Earth.

Six humans living the lifestyle of the average North American wouldn't pose much of a problem if they were the only people on Earth. However, there are currently about six *billion* humans on Earth and this number continues to increase. It is anticipated that the human population will top out at about 12 billion souls roughly 100 years from now. This means twice as many people as today living off of Earth's limited resources. All else being equal, the implication is that humanity's impact on the environment may double by that time. Therefore, reducing the burden humans place on the environment requires a two-pronged strategy: on the one hand, reduce the damage caused by each person, and, on the other, curb global population growth.

What You Should Do

Extrapolation and Its Limitations

Science is often expected to make predictions about the future from information available today. For example, the global population of humans 50 years from now might be expected to be predictable from current censuses and growth rates worldwide. Such an expectation is certainly legitimate. Unfortunately, available data is often limited in quantity, incomplete, or uncertain, seriously compromising the reliability with which science is able to yield predictions.

An attempt to predict the future begins with a graph of data versus time. The points define a path (or **trend**) which can be extended into the future in order to make predictions for data in the future. This procedure is called **extrapolation**. If the path followed by the points is approximated to be that of a straight line, then the procedure is called **linear extrapolation**.

However it is done, extrapolation is dangerous. It can lead to misleading results if the data don't continue along the adopted path into the future (say, if circumstances are changing), or if the data don't really follow the adopted path to begin with (even though points may appear to do so due to failings of the data).

To explore the process of extrapolation, work through the following exercise:

1. Using the first piece of graph paper provided with this activity, plot the data for the *first three* times listed in Table 1 and then stop. Use different colours for each set of objects (cakes, fast bacteria, and slow bacteria).

2. Imagine that you don't have the rest of Table 1. For each set of objects, draw a single straight line that best tracks the three data points (note that the points won't necessarily lie on the line). Extrapolate the line to make a prediction. **Based on the data for the first three times alone**, how many cakes do you expect will have been made by 10 PM? How many fast and slow bacteria do you expect there to be at that time?

3. Complete the plots with the remaining data in the Table. How reliable were your extrapolations?

Your Growing Family

In many ways, the growth in the population of your family mimics that of a country, with marriages or adoptions akin to immigration, and divorces akin to emigration. To explore and illustrate the power of exponential population growth, chart the increase in your own family's numbers. First, on your log sheet, create a family tree going back two generations, i.e., starting with your grandparents at the very top. Indicate the years of birth and death (if appropriate) for each person, and also indicate the year in which any marriages or divorces took place.

Next, using the information from your family tree, create a *chronologically-ordered* table of significant events for your family. A significant event is one where a person was added or removed from your family, i.e. birth, death, marriage, divorce, adoption, etc. For each event, record in the table how many persons were added or removed. For each year in which at least one event occurs, calculate the total number of persons in your family in that year starting from the previous total. Add the number of persons who joined your family and subtract the number of persons who left your family. For example, say the total number of persons in your family in 1965 was 23, but then in 1967 there were two births, one death, and one marriage. The total population in 1967 would be $23 + 2 - 1 + 1 = 25$.

To better understand your data, prepare a graph of your family's growth using the second sheet of graph paper provided. Leave enough room to the right of your data on the graph to extrapolate out to future generations. The year should be on the horizontal axis and the total number of individuals in your family should be on the vertical axis.

Using your graph, answer the following questions: (Draw any paths you define for extrapolation using a different colour.)

1. Do you notice any trends? Is the growth of your family linear? If not, then how is the population growing?

2. How many individuals will your family include in the year 2100? On the graph, clearly mark and label any path you use to extrapolate.

3. Judge the uncertainty in your estimate of your family's population in the year 2100.

What You Will Need

From Kit

- 2 sheets of graph paper (tear-outs included with this activity).

Things You Should Think About

- On the graph displaying your family's population growth, you should see clusters of events at regular intervals. For example, a burst in numbers will occur when individuals marry and have children. The separation of clusters defines the length of a **generation**. See if you can recognize the onset of new generations in your family. Due to healthier living, the length in years of a generation may change with time. Do you see evidence for this?

What You Should Disseminate

Page 1
The graph exploring extrapolation and its limitations as applied to cakes and bacteria.

Page 2
Summarize your explorations and findings about your family, including your answers to the questions about it.

Page 3
The graph of your family's growth.

Page 4
Your original log sheet.

Log Sheet: Exponential Growth

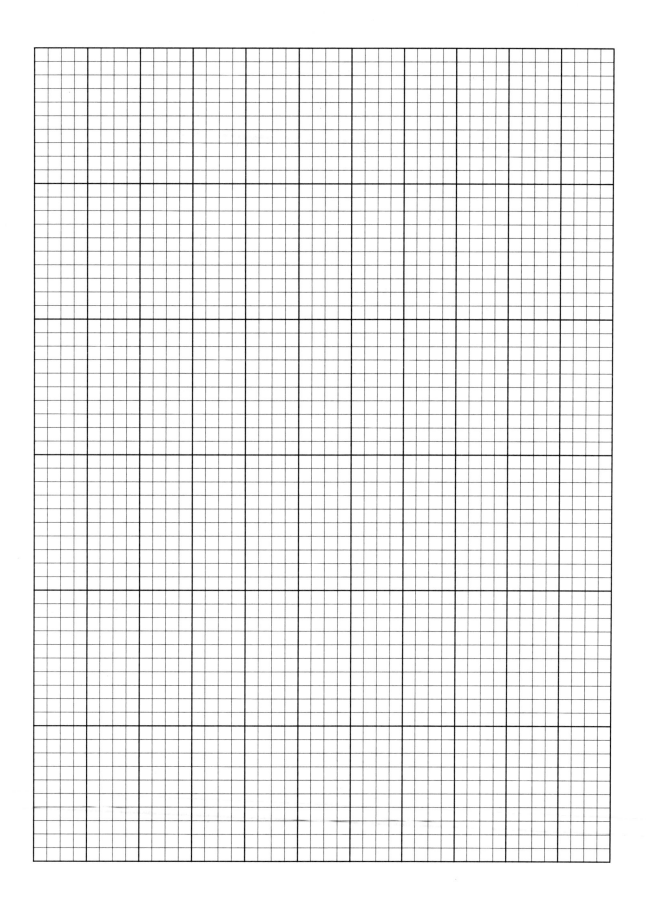

Activity 2 Overpopulation

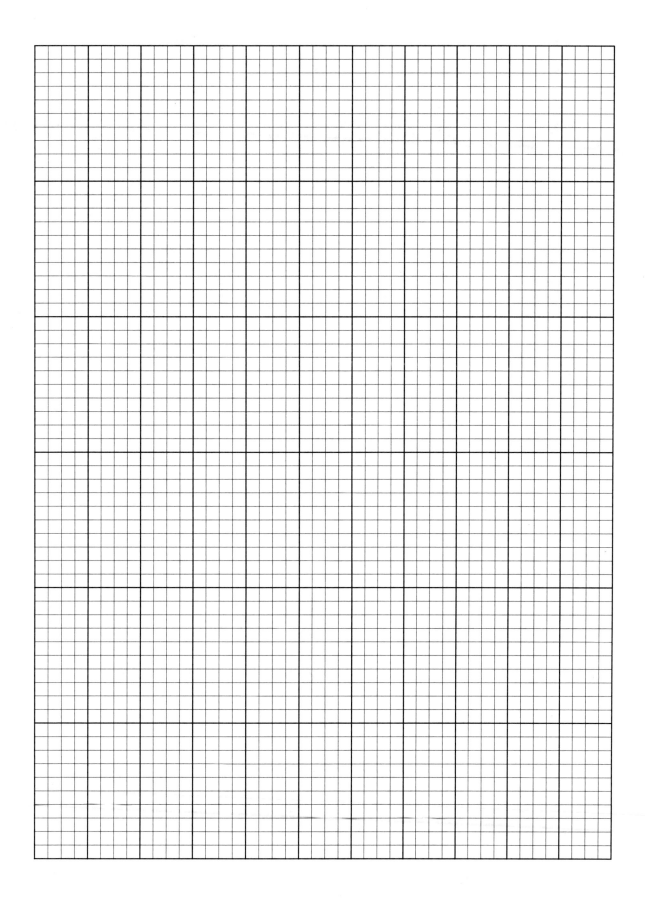

Activity 3
Global Warming
The Enhanced Greenhouse Effect

What Is Temperature?

Global warming is all about changing temperature. To understand it, you need to know what temperature is. When something is heated, its constituent atoms and molecules are stimulated to move around faster. This is because energy is absorbed from the heat source. When you place a thermometer into air the atoms and molecules of the air bump up against the atoms and molecules of the thermometer. In other words, the constituents of the thermometer *feel* the constituents of the air, and they end up moving around at a speed consistent with that of the constituents of the air. The accompanying visible response, which in the case of a bulb thermometer is the expansion of the liquid in the bulb, is translated into a number we call the temperature. **Temperature** is simply a statement about how fast atoms and molecules are moving.

Temperature is determined by the environment. For example, a rock in the shade is warm because atoms and molecules in the air continually run into it. A rock in the Sun gains additional warmth by absorbing some of the energy contained in sunlight.

An important consequence of temperature is that everything glows. This is because atoms and molecules continually bump into each other, often causing them to take in energy from each other and become excited. Molecules, for example, display their excitement by spinning or vibrating faster. An atom or molecule which is excited eventually returns to a calmer state by releasing its excess energy as an **electromagnetic wave**, better known as a **light wave**. The combined effect of many atoms and molecules becoming calmer is a glow.

At night, things like rocks look dark, but in fact they are not. The glow of light from their warmth can't be seen, because the colour of the light is **infrared** (beyond the red), to which our eyes are not sensitive. *When you raise the temperature, not only do atoms and molecules move faster, but the rate at which infrared light waves are emitted increases.* Note that in the daytime things look bright only because they are reflecting sunlight. This is *not* the glow associated with their temperature.

What Is a Steady State?

Warming is a sign of change. Yet, nature prefers a **steady state**, which is one where the temperature is not changing. The temperature of any body stays fixed as long as the rate at which energy is emitted (cooling) is balanced by the rate at which energy is absorbed (heating). What might cause temperature to change? Suppose a rock is moved from the shade into sunlight. Suddenly the rock sees more energy hitting it every second (the sunlight). Atoms and molecules in the rock take in some of this energy (as against reflecting it away), so the rock starts warming up. However, the rock doesn't like change, and desperately seeks a steady state. Ultimately, it becomes so warm that the rate at which it emits energy once again balances the rate at which it is absorbing energy from its environment. Happily, the rock's temperature stops changing. A steady state has been achieved, but the rock is now warmer. In summary, the reason something changes temperature is to remove an imbalance between cooling and heating rates.

What Is the Greenhouse Effect?

The atmosphere of the Earth acts like a blanket, trapping infrared waves coming from below. This leads to a warm Earth, a phenomenon known as the **greenhouse effect**. Here are the details.

The atmosphere admits light from the Sun, which then heats the ground and air. The heated ground and air emit infrared light waves, many of which head upward toward space. However, certain gases in the air are composed of molecules which are easily excited when exposed to infrared waves. They absorb many of the waves, the energy from which is converted into faster vibrations. Atmospheric gases which are particularly adept at blocking infrared waves are called **greenhouse gases**. Examples are carbon dioxide, methane, and water vapour. Excited molecules eventually return to calmer states by emitting infrared waves. Some of these waves come back down through the atmosphere to the ground, heating both the air and the ground further. Thus, to reach a steady state where the rate of absorption of sunlight energy is balanced by the rate of escape of energy to outer space, the Earth has to boost the rate at which it emits infrared waves. This is to compensate for the fact that not all of the infrared waves emitted are able to escape. Consequently, in a steady state the Earth has to be warmer than would be the case if there were no atmosphere.

What Is the Enhanced Greenhouse Effect and Why Should You Care?

As long as the Earth has had an atmosphere, there has been a greenhouse effect. It is one of the major reasons why the Earth is a habitable planet. With it, the average surface temperature is 20° C. Without it, the average temperature would be more than 20° lower, namely below freezing. Of course, there are times everywhere when the temperature is far below the average, so it is questionable as to whether life could survive on such a cold world.

The greenhouse effect in and of itself is a blessing, not a concern. What is important to us now is that the greenhouse effect is becoming stronger, in turn causing the whole planet to get warmer. This is because of growing quantities of carbon dioxide gas. The most likely cause is the burning of fossil fuels. The Earth wants to be in a steady state; it is absorbing energy from the Sun at roughly the same rate as always, but can't emit energy as fast because more of what it is trying to emit is trapped. Consequently, it is forced to get warmer, to boost the rate of emission to the point that it balances the rate of absorption. This is **global warming**, and it is rooted in the **enhanced greenhouse effect**.

How Might You Measure an Enhanced Greenhouse Effect?

To observe an enhanced greenhouse effect, all you have to do is to create a "greenhouse" in which the air temperature is raised above the ambient value by the trapping of infrared light waves. This means you have to introduce a walled-off zone where sunlight is admitted through the walls, but infrared waves generated by the heating of the walls and the air or the ground inside are blocked by the walls. Glass or clear plastic is a good choice for the walls, because the molecules of which they are made behave just like the molecules in the Earth's atmosphere. To observe the enhanced greenhouse effect, then, all you have to do is to place a glass or clear plastic container in sunlight (direct or indirect, although the former works better), wait a while (say half an hour to an hour) to allow the container to reach a new steady-state temperature, and then measure the temperature of the air inside. Note that it is appropriate to describe the effect as an enhanced one because the container walls are enhancing the greenhouse effect already set up by the Earth's atmosphere.

It is likely that you have already observed the enhanced greenhouse effect. When you park a car outdoors, the windows admit light from the Sun but block infrared waves from the heated interior from leaving. Thus, the air inside of the car becomes warmer than the air outside.

What You Should Do

Design and execute a useful experiment to test the following hypothesis:

> As a result of the enhanced greenhouse effect, the air in a sealed container with transparent walls becomes warmer when exposed to sunlight.

Consider using dried soil to define the "ground". If there are chunks, you should try to break them up. Records should be entered in pen on the included log sheet.

Your thermometers are fragile – be careful not to break them, e.g., don't try to bend them.

What You Will Need

From Kit

- Thermometers

From Other Sources

- One or more glass or plastic containers which can be sealed (you decide how many)
- Dried granular soil

Things You Should Think About

- A truly useful experiment is one which is controlled, statistically significant, and repeatable. In the case of a car, would it become just as warm if it had no windows? To tell with certainty if the windows make any difference you would have to perform an experiment to compare the temperatures inside cars with windows to the temperatures of identical cars with the windows open.

- The size of the enhanced greenhouse effect, and thus its measurability, will depend upon how well your walls block infrared light waves from escaping from the inside of your container. Thick-walled containers should trap heat better than thin-walled containers. In essence, going to a thicker-walled container is akin to injecting more molecules of carbon dioxide into the atmosphere of the Earth.

- Beware of other variables. For example, due to a property known as the albedo, which is discussed in another activity, the ground inside and outside your container may have a bearing on the temperature you measure.

- In measuring the enhanced greenhouse effect, you don't really care what the actual temperature is. What you care about is how the temperature in your "greenhouse" compares with that of a control.

- Of some concern might be differences in construction of different thermometers. Owing to differences in construction, different thermometers may give different readings even if they are all at the same temperature. You must take care to study this issue, and take measures to deal with any problem while doing your experiment.

What You Should Disseminate

Page 1
Describe your experiment and your findings.

Page 2
Show a detailed drawing of your experimental setup. It should be carefully and liberally labeled.

Page 3
Your original log sheet.

The first page should include:

- A statement of the hypothesis being tested.

- A description of the experiment you designed to test the hypothesis.

- A description of where, when, and how you made the measurements, including uncertainties.

- Your conclusions, especially with reference to the hypothesis.

Log Sheet: The Enhanced Greenhouse Effect

Activity 4
Global Refrigeration
The Albedo

What Is the Albedo?

In studying the enhanced greenhouse effect, you learned about the concepts of temperature and a steady state. How warm an object that is exposed to light has to get to reach a steady state depends upon how well it absorbs the energy carried by light waves. This depends upon what it is made of. A rock which absorbs a large fraction of the sunlight incident upon it (reflecting very little away) will have to end up emitting infrared light waves at a high rate to achieve a steady state. This means it must become hot. However, a rock which absorbs a small fraction of the sunlight incident upon it (instead reflecting most away) need not emit infrared waves so rapidly to strike a balance, so in its steady state it will be cooler.

The **albedo** is a quantitative statement about how well an object reflects light. Since what is not reflected is absorbed, it also conveys information about how the temperature might be expected to adjust as a result of increased exposure to light. It is defined as follows:

$$\text{Albedo} = \frac{\text{Rate at which light energy is reflected}}{\text{Rate at which light energy arrives}}$$

In words, the albedo is the fraction of the light arriving which is reflected. It has a value less than 1.0, since nothing can reflect more light than it receives. Note that the albedo is not dependent upon how big the object is. You can define it for a small piece of a rock, a whole rock, or even a whole planet simply by comparing the rate at which energy is coming in to the rate at which it is reflected out.

An object with a high albedo is one which is highly reflective, and thus a poor absorber. An object with a low albedo is one which does not reflect well, and thus is a good absorber. If you were to move a rock with a high albedo and a rock with a low albedo from the shade into sunlight, you would find that the rock with the high albedo would end up being the cooler. This is why the albedo is often referred to as a **refrigerant**. All other things being equal, the higher the albedo, the greater the *refrigeration*.

How Does the Albedo Affect Climate?

Planets are continually bathed in light from the Sun. How warm they get, i.e., their climate, depends in part upon the efficiency with which they absorb the energy in the sunlight. That is determined by the albedo. In essence, the albedo regulates the production of infrared waves (the refrigeration). The greenhouse effect determines the fraction which are trapped (the warming). The two factors together determine climate.

Of course, the albedo varies from place to place around a planet, depending upon what is in the way of the sunlight. The albedo is high where there are clouds. It is high where there is ice. It tends to be lower where there are forests, and can be lower still if there is exposed ground. The albedo for a whole planet depends upon how all these things are balanced.

In the case of the Earth, global warming appears to be causing the polar ice caps to recede. With ice covering a smaller fraction of the globe, the Earth ought to become less reflective overall and the albedo should drop. With a lower albedo, the Earth would be more adept at absorbing sunlight and would get warmer still. In turn, ice would melt more rapidly and the albedo would decrease further. This is rather disconcerting, because it means that the warming trend we see today might get amplified by this **feedback loop**.

How Might You Measure the Albedo's Effect?

To measure the effect of the albedo on climate, you need to create an environment which mimics that of a planet. For example, a sealed jar would do the trick. What you put on the bottom of the jar would be the "ground." The walls would act like additional greenhouse gases, admitting sunlight but increasing the blocking of infrared light waves emitted by the ground and air inside (the enhanced greenhouse effect). The lid would ensure that the the contents of the jar would achieve a steady state decoupled from whatever state your laboratory (the Earth) is in.

All you have to do to measure the effect of the albedo on climate is to examine how different ground covers influence the temperature (the climate) which is achieved when the jar is illuminated with sunlight (direct or indirect, although direct should work better). You just have to be sure to wait long enough, say an hour, for the climate inside the jar to reach a steady state before measuring the temperature.

The albedo of the ground controls the rate at which sunlight entering the jar is reflected. Reflected sunlight largely escapes without heating the contents of the jar. What is not reflected is absorbed by the ground and re-processed into infrared waves. Since these waves can't readily escape the jar, they contribute to the warming of the contents. The balance between reflection and absorption determines the steady-state temperature ultimately achieved by the jar as a result of the enhanced greenhouse effect.

What You Should Do

Design and execute an experiment to test the following hypothesis:

The albedo of a planet has an impact upon climate.

Do so by comparing the effect of different ground covers on the climate inside a transparent container. Use dried topsoil (grind up any chunks) to represent exposed ground. Use salt, which is included in your kit, to represent snow or ice. Also experiment with another ground covering from whatever materials you have around your home. Records should be entered in pen on the included log sheet.

What You Will Need

From Kit

- Thermometers
- Salt

From Other Sources

- One or more glass or plastic containers which can be sealed (you decide how many)
- Dried granular topsoil
- A ground cover of your choosing

Things You Should Think About

- You will not be able to measure the actual albedo of any given ground cover. However, you can compare albedos.

- The sensitivity of the steady-state temperature to the albedo of the ground will depend upon how well your walls block infrared light waves from escaping. Thick-walled containers should trap infrared waves better than thin-walled containers.

- In order to allow your ground to intercept as much sunlight as possible, and thereby influence climate maximally, your ground cover should be spread out over as large an area as possible. Be sure to break up any chunks. Consider turning containers on their side.

- Beware of other variables. For example, you should ensure that the enhancement of the greenhouse effect by all containers you employ is the same, and that they are all illuminated equally.

- When damp soil is heated, water will evaporate. Water vapour is adept at blocking infrared waves, and may influence climate by enhancing the greenhouse effect further. Avoid damp soil.

- Of some concern might be differences in construction of different thermometers. Owing to differences in construction, different thermometers may give different readings even if they are all at the same temperature. You must take care to study this issue, and take measures to deal with any problem while doing your experiment.

What You Should Disseminate

Page 1
Describe your experiment and your findings.

Page 2
Show a detailed drawing of your experimental setup. It should be carefully and liberally labeled.

Page 3
Your original log sheet.

Page 1 should include:

- A statement of the hypothesis being tested.

Activity 4   Global Refrigeration

- A description of the experiment you designed to test the hypothesis.

- A description of where, when, and how you made the measurements, including uncertainties.

- Your conclusions, especially with reference to the hypothesis. On the basis of your experimental results, rank the albedos of the various ground covers you have evaluated from lowest to highest.

Log Sheet: The Albedo

Activity 5
Ozone Depletion
Ultraviolet Light

What Is Ozone Depletion?

Most of the light received by Earth from the Sun comes in the form of visible light. But the Sun also delivers significant amounts of light not visible to our eyes, such as ultraviolet (UV) light. Large amounts of UV light potentially pose a threat to life on the surface of the Earth. Fortunately, the layer of **ozone** molecules that lies at an altitude of between 17 and 48 km above the ground, in a part of the atmosphere called the **stratosphere**, filters out most of the ultraviolet light from the Sun. But it wasn't always there. Indeed it is believed that life on Earth originated in the oceans, where it was shielded from UV light by water, and was unable to emerge from the oceans onto land until the protective ozone layer had formed.

In 1985, British scientists reported an **ozone hole** above Antarctica, namely a huge zone of the atmosphere where the amount of ozone had diminished. Subsequently, scientists found a similar zone above the North Pole. It is now recognized that stratospheric ozone levels have gone down everywhere, a phenomenon known as **ozone depletion**. However, the greatest losses have occurred above the Earth's poles, which is why the problem was first discovered there. The thinning of the ozone layer has increased the exposure of living things to UV light.

In the 1970s, it was discovered that certain chemicals known as chlorofluorocarbons (CFCs for short) destroy ozone when released into the atmosphere. Such chemicals were commonly used as refrigerants or as propellants in spray cans. It was concluded that ozone depletion was caused by humans. Since then, steps have been taken to restrict or ban the use of CFCs. Contaminants already in the air are transforming (naturally) into less harmful forms, but it will be many decades before ozone levels return to normal.

What Is Ultraviolet Light?

Just like visible and infrared light, **ultraviolet light** is a type of electromagnetic wave. These three types of electromagnetic waves are distinguished from one another by their **wavelength**, defined as the distance between one wave peak and the next. The wavelength defines the colour. Electromagnetic waves form a continuous **spectrum** from very, very short wavelengths to very, very long wavelengths. Visible light waves, which our eyes can detect, have wavelengths from 4×10^{-7} to 7×10^{-7} metres (blue to red). Infrared waves, which are associated with heat, have longer wavelengths, up to about 1 millimetre. Ultraviolet waves have shorter wavelengths, down to 10^{-8} metres.

The UV part of light is usually itself divided into three parts that are distinguished by wavelength. So-called UV-C light, which has the shortest wavelengths, is the most biologically damaging. Fortunately, the Earth's atmosphere, and in particular the stratospheric ozone layer, filters out almost all of the UV-C light from the Sun. UV-A light has the longest wavelengths in the UV region and is much less harmful, though exposure to it leads to tanning and ageing of skin. Because of its longer wavelengths, UV-A passes through the atmosphere largely unhindered and constitutes 99% of the UV light reaching the Earth's surface. UV-A light also passes through glass. Lying in between the UV-A and UV-C extremes is UV-B light. UV-B waves are responsible for sunburn. It is UV-B light which is most affected by ozone variability.

Why Is UV Light Harmful?

Why UV light is harmful while visible light isn't is as follows: a wave with a shorter wavelength packs more energy. So, UV waves always carry more energy than infrared or visible waves, just as visible waves always carry more energy than infrared waves. More specifically, UV waves carry enough energy to damage living cells and their constituents, such as DNA molecules. This is especially true of UV-B and UV-C waves.

A suntan is the skin's way of trying to protect itself from UV waves. When exposed to UV light, the body produces melanin, a pigment that readily absorbs the energy of UV waves. However, melanin offers only limited protection, roughly that of a suntan lotion with a Sun Protection Factor (SPF) of 4. The SPF rating describes how much longer it takes for protected skin to burn compared to unprotected skin. With an SPF of 4, skin that otherwise starts to redden within 10 minutes takes about 40 minutes, that is four times longer, to reach the same degree of reddening. But regardless of the SPF, when the skin's defenses are overwhelmed, the result is a sunburn. In this situation, the energy from UV waves is deposited into skin cells, damaging them or their constituents. Genetic damage can be passed on to successive generations of cells. Cancer is often the end result of this process.

The ozone layer provides some protection against UV waves, but as noted above it has recently been eroded. Even though the manufacture and use of ozone-destroying chemicals has been prohibited by international treaties since the early 1990s, it is believed that it will take at least fifty years for the quantities that are already in the atmosphere to lose potency and for the ozone layer to recover. This means that we are likely to be exposed to greater amounts of UV light for some time. It is therefore important for each person to know under what conditions UV light is most intense and also how to protect him/herself, i.e. what materials do and don't block UV light.

How Might You Detect UV Light?

Certain molecules come in two (or more) different configurations. In an aggregate, a particular configuration is manifested by some physical characteristic, such as colour. Thus, a substance composed of such molecules should change colour if the configuration of each molecule changes. One way to force a change in a molecule's configuration is to deliver energy, such as that carried by a UV wave.

The plastic beads included in your kit have been impregnated with a substance that has precisely the property described above. When exposed to UV light they change colour. In the absence of UV light they revert back to their original white within a short time.

What You Should Do

Sources of Light

Become acquainted with how your UV-sensitive beads respond to different lighting conditions. To make it easier to recognize subtle changes in colour, and to prevent colour bias which might arise from varying backgrounds, two colours of beads are provided. The beads that turn orange respond more slowly to UV light than those that become purple. Try to establish some sort of scale with which to describe the intensity of the colour, ranging from no colour to very vivid colour.

Explore the strength of the response of the beads (change in intensity of the colour) when exposed to

1. Direct sunlight
2. Cloudy conditions
3. Ordinary light bulb
4. Fluorescent light
5. Halogen light (if available)

Blocking UV Light

Having determined that the beads respond to invisible light from the Sun (which we are telling you is ultraviolet), investigate ways to block this light. Try the following materials:

1. Glass
2. Plastic
3. Water (of varying depths)
4. Sunglasses (or a suitable substitute)

You can also try other materials of your own choosing.

Sunscreens

For sunscreens, it is possible to go one step further and to use the beads to explore the effectiveness of different Sun Protection Factors. So, design and execute a useful experiment to test the following hypothesis:

Sunscreens with a higher SPF are more effective at blocking UV light than those with a lower SPF.

What You Will Need

From Kit
- UV-sensitive (colour-changing) beads
- Clear plastic bags
- Opaque plastic bag
- Sunscreen samples
- Beading string (to make your very own colour-changing bracelet when you're done)

From Other Sources
- UV-blocking sunglasses (if available)

Things You Should Think About

- It is a good idea to slip a piece of white paper under the beads to enhance the contrast between the beads and the background.

- The beads will float in water, so to immerse them you may have to weigh them down with something.

- Does the beads' response to sunlight depend on the time of day?

- To test sunscreen, consider putting some beads into a plastic bag (from the kit) and coating the outside of the bag with sunscreen.

What You Should Disseminate

Page 1 and 2
Describe your explorations, experiments, and findings.

Page 3
Provide a drawing of your experimental setup for testing sunscreen to clarify how you performed your measurements.

Page 4
Your original log sheet.

Pages 1 and 2 should include:

- A summary of your investigations into how the beads respond to different light sources.

- A summary of your investigations into how the beads respond when different materials are employed to block UV light.

- A statement of the sunscreen hypothesis.

- A brief description of the experiment you designed to test the hypothesis.

- A description of where, when, and how you made the measurements. Be sure to note the weather conditions when relevant.

- Your conclusions, especially with reference to the hypothesis. Point out any surprises or unexpected results and attempt to explain them.

Log Sheet: Ultraviolet Light

Activity 6
Energy
Solar Power

What Is Energy?

The **energy** of a system is the ability of the system to do work. **Work** is done whenever a mass moves in response to a force acting on it. Energy can take on many forms:

- When a cannon ball collides with a wall, work is done on the wall causing damage. We say that a moving object carries **kinetic energy**, that is energy associated with its motion.

- Opposite electrical charges which are held apart from one another have the potential to come back together if they are released, because an attractive electromagnetic force acts between them. Thus, a system of separated charges can be said to have **electrical potential energy**.

In general, **potential energy** refers to energy that is stored in a physical system and that can be converted to other forms of energy, such as kinetic energy, under the right circumstances. Note that nature always strives to find a way to minimize potential energy, say by bringing unlike charges closer together.

Whereas energy is the ability to do work, **power** is the rate at which energy is delivered to or extracted from a system. Energy is expressed using a unit called a **Joule** (J), and power is expressed using the **Watt** (W). One Watt is equal to one Joule every second. For example, a 60 W light bulb consumes 60 J of electrical energy every second it is on. If it is on for one minute, then it consumes $60 \text{ J/s} \times 60 \text{s} = 3600$ J of energy in total during that time.

Two laws of nature apply to energy. First, in any physical process energy can neither be created nor destroyed. It can only be transformed from one form to another. Second, in any transformation there is always a degradation of some of the energy into a less useful form, such as heat. The first law means that the only energy available to us is what we receive from space (mostly the Sun) plus whatever is already stored on Earth.

The human race makes use of all forms of energy, which we obtain from several different sources. Some of these, such as fossil fuels and uranium, are **non-renewable**, which means that we will eventually run out of them. On the other hand, **renewable** energy sources, such as flowing water, wind, and solar energy, are expected to last as long as the Earth.

What Is Solar Energy?

The Sun is primarily composed of hydrogen, the simplest and most abundant element in the universe. In the Sun's core, the temperature and pressure are so great that hydrogen nuclei can fuse together to form helium. This process, known as **thermonuclear fusion**, releases a tremendous amount of energy. The internal nuclear furnace keeps the surface of the Sun at a temperature of about 6000° C. The Sun radiates a huge amount of energy in the form of **electromagnetic (EM) radiation** that, at this temperature, is concentrated in the *visible* part of the spectrum. In other words, energy is emitted as light that we can see. The Sun's total energy output every second, i.e., its power output, is about 3.8×10^{26} Watts. The power emitted from just one square metre of the Sun's surface is a mere 62 million Watts!

As one moves away from the Sun, the power emitted 'spreads out'. Thus, the power crossing a one square metre area decreases rapidly with distance from the Sun. By the time sunlight reaches the top of the Earth's atmosphere (about 8 minutes after escaping the Sun's surface), it has spread out so much that it delivers a comparatively modest 1366 Watts per square metre (W/m^2). Besides making the Earth habitable, solar power provides all of the energy for photosynthesis in plants, which are at the base of the food web, and is the origin of almost all of the renewable energy sources on the planet, including wind, hydroelectricity, and solar electricity. It is expected that solar power will be available for at least another five billion years.

The solar power incident on a surface is called **insolation**. (Don't confuse this with *insulation*, which is a material used to reduce heat loss from a home (for example).) Insolation is usually expressed as the incident power divided by the area of the surface, i.e. in W/m^2. It depends on the angle between the surface and the Sun's rays, as can be seen by thinking about the 'shadow' cast by the surface. Figure 3 shows a surface at two different angles. The shadow results from the interception of light by the surface, so it follows that a larger shadow corresponds to greater insolation. The size of the shadow is a maximum when the surface is perpendicular to the Sun's rays (i.e. directly facing the Sun), which means that insolation is also a maximum then.

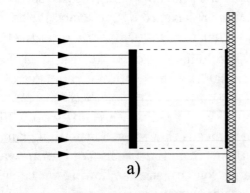 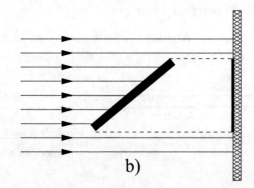

Figure 3: Energy intercepted by a surface. In a) the surface directly faces the incident light and the size of the shadow cast on a wall (shown by the dark vertical line) is a maximum. In b) the surface has been tilted with respect to the direction of the incident light, and the shadow it casts is smaller.

What Is the Insolation At Earth's Surface?

The insolation of a horizontal surface at some location on Earth at a given time depends only on the elevation of the Sun in the sky at that time. The **elevation** (how high the Sun is in the sky) comes into play for two reasons.

The first has to do with the angle between the surface and the Sun's rays. If the Sun isn't directly overhead, then a horizontal surface is at an angle to the Sun's rays and, as just discussed, the insolation of that surface is reduced.

The second reason for which insolation depends on elevation has to do with the Earth's atmosphere. Sunlight gets scattered or absorbed by air molecules. These effects reduce surface insolation. Suppose that the Sun is directly overhead. By the time sunlight reaches the surface of the Earth, the insolation of 1366 W/m^2 at the top of the atmosphere has dropped to about 1000 W/m^2. But, as illustrated in Figure 4, if the Sun is closer to the horizon (i.e., at a lower elevation), then the insolation is even less because sunlight has to penetrate a thicker layer of atmosphere. More light is lost to scattering and absorption.

Combining both effects, it is easy to see that insolation is significantly lower at sunrise and sunset than it is at midday. It also follows that average daily insolation decreases as one moves north or south from the equator to higher latitudes. At higher latitudes, the Sun's maximum elevation during the day is lower than it is at the equator. Thus, averaged over a whole day, insolation is necessarily less.

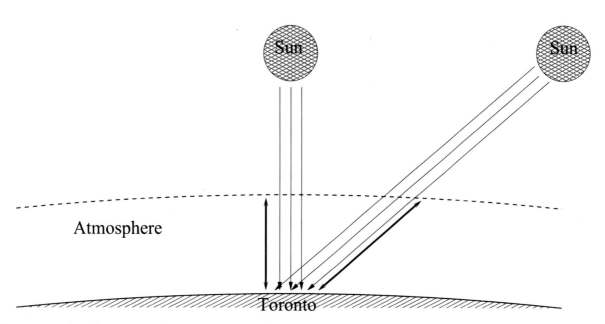

Figure 4: Insolation at Toronto for two different Sun elevations. When the Sun is directly overhead, the path length of light through the atmosphere is a minimum. When the Sun is closer to the horizon, the path length is longer.

Activity 6   Energy

What Is Electricity?

Electricity refers to electrical energy. Any electrical charge produces what is called an **electrical potential field** whose value depends on distance from the charge. Charges of opposite sign which are held apart, as in a battery, create a varying electrical potential field between them. How much the field changes from one side to the other is called the **electrical potential difference**. Any additional electrical charges placed in the field, such as those in a wire connected to the terminals of a battery, will be set into motion in a way that serves to lower electrical potential energy – they move toward the charges of opposite sign. The resulting kinetic energy of the moving charges, which is determined by the electrical potential difference, may then be harnessed by an electrical device to do some sort of useful work. Electrical potential difference is usually expressed in **Volts**, and so it is also often simply called **voltage**.

To obtain electricity there must be some mechanism to produce and maintain a charge separation. Batteries and generators do this. A battery is a device specifically designed to maintain a certain electrical potential difference between its poles by chemical means (eg., a 1.5 V battery). If a connection is made between the poles of a battery using a metal wire, then negatively-charged electrons in the wire will be drawn to the positive pole of the battery. A flow of charge, i.e., an **electrical current**, results. Electrical current is expressed in terms of a unit called the ampere. One **ampere** (1 A) is a flow past some boundary of 6.2×10^{18} electrons every second – it is a comparatively large current. In many everyday applications, electrical current is more conveniently expressed in milli-amperes (mA), where one milli-ampere is one thousandth of an ampere.

Traditionally, electricity has been produced using a battery, which converts chemical energy into electrical energy, or by using a generator, which converts kinetic energy into electrical energy. More recently, it has become viable to produce electricity directly from sunlight using **solar cells**.

What Are Solar Cells?

Visible light waves with a short enough wavelength carry enough energy to knock electrons out of some metals. This is called the **photoelectric effect**. The 1923 Nobel Prize in Physics was awarded to Albert Einstein for his successful explanation of this effect: that light waves can behave like particles. Thus, a single light wave is referred to as a **photon**.

Solar cells are made of a semiconducting material (such as silicon) to which specific impurities are added. When hit by a photon of sufficient energy, an electron is freed and moves to one face of the solar cell. Thus, whenever the cell is exposed to light, an electrical potential difference develops between the two faces of the cell. A solar cell behaves just like a battery – a battery that will never run down – as long as light is present. So if an electrical device is connected to both faces of a solar cell, then an electrical current flows from which energy can be extracted to do useful work. The speed at which electrons flow is determined by the electrical potential difference, which depends on the materials from which the solar cell is made. The size of the current is determined by the brightness of the light. If the cell produces enough current, it could turn on a light bulb. The energy emitted by that light bulb started out as sunlight.

Suppose again that the Sun is directly overhead and that the insolation is 1000 W/m^2. Does that mean that a one square meter solar panel will produce 1000 W of electricity? Sadly, the answer is no. Typically, only 14% to 20% of the photons hitting a solar cell end up freeing electrons and thereby contributing to the maintenance of an electrical potential difference. With today's cells, one might obtain only 140 W to 200 W of electricity from 1000 W of sunlight. Reasons for the low conversion efficiency are many and include:

- Panel albedo – any incident sunlight that is reflected by the solar panel doesn't get converted to electricity. Reflections are unavoidable with real materials.

- Spectral sensitivity – a solar cell is typically only capable of converting a rather narrow range of wavelengths of light into electricity. Other wavelengths can be absorbed by the cell but are not converted to electricity.

Efforts to improve solar cell efficiency are focussing on broadening spectral sensitivity and improving anti-reflection coatings (among other things). Such improvements are essential to promote wider adoption of solar electricity generation.

What You Should Do

Your kit includes a solar cell and an ammeter, as well as a pair of wires which should be used to connect them to each other. An **ammeter** measures electrical current, so it can be used to detect the electricity generated by the solar cell. To assemble a solar power detector, first connect the positive (+) terminal at the back of the solar cell to the positive (+) terminal at the back of the meter. Then, connect the negative (−) solar cell terminal to the negative (unmarked) terminal on the meter.

Now, if the cell is exposed to sufficiently bright light, the ammeter needle will move clockwise by an amount that reflects the current flowing in the circuit (in milli-amperes). (If you find that the needle tries to go in a counter-clockwise direction, then you've connected the wires incorrectly. Reverse your connections and try again.) The solar cell provided will develop a potential difference of up to 0.45 V when it is exposed to an insolation of 1000 W/m^2. This will lead to a current of up to 400 mA.

Once you have connected the circuit, cover the solar cell completely and use a small screwdriver to turn the adjustment screw on the front of the meter until the needle reads zero (as it should for zero incident light). This calibration step is important to avoid systematically incorrect readings and should be done prior to making any measurements.

Light Sources

The Sun isn't the only source of electromagnetic waves. So, first explore the ability of different light sources to generate a current in your circuit. Hold the solar cell right next to the light source and measure the generated current. How far away from each source can you get before the current drops to zero? If possible, you might also explore how the current depends on the wattage of the bulb.

Activity 6   Energy

Insolation

The insolation of a solar cell should depend on the angle it presents to the Sun and on the amount of air that lies between it and the Sun. Design and execute a repeatable experiment to test the following two hypotheses:

1. The insolation of a surface is greatest when it directly faces the Sun.

2. The insolation at the surface of the Earth is reduced when light passes through more air.

You will want to measure and graph how the insolation (as indicated by the current) depends on the tilt of your solar cell. So, you will need a way of expressing the angle between the surface of the solar cell and the Sun's rays. If you place the solar cell on a flat horizontal surface in direct sunlight and you tilt it towards the Sun, then it casts a shadow. It turns out that the length of that shadow is a good proxy for the angle of interest.[1]

What You Will Need

From Kit

- Solar cell
- Ammeter
- 2 wires with alligator clips on the ends
- Ruler
- 1 sheet of graph paper (tear-out included with this activity)

From Other Sources

- Light sources

Things You Should Think About

- How do you know when the solar cell is directly facing the Sun?
- If the *edge* of the solar cell faces the Sun, then the shadow cast by it is a minimum. Tilting the cell away from this edge-on orientation produces a shadow on one side or the other of the cell, depending on the direction of the tilt. How might you distinguish quantitatively between the shadows cast in either direction?
- Do you see a difference in current when the solar cell is pointing skyward (but not directly facing the Sun) compared to when it is pointing 'earthward'? Does this difference (if any) make sense?
- How might your measurements change if you were to repeat the experiment on a cloudy day?
- If you have to re-calibrate your meter during the experiment, is it appropriate to compare results taken before with those taken after?
- How do you think the peak (midday)

---

[1] For the mathematically-inclined, the length of the shadow is directly proportional to the cosine of the angle between the direction to the Sun and the direction perpendicular to the surface of the solar cell.

current changes at your location at different times of the year?

- Suppose you have purchased a solar panel for your home. What would be the best orientation for it, supposing that it will be fixed forever?

What You Should Disseminate

Page 1
Describe your experiments and your findings.

Page 2
Graphs showing current as a function of shadow length.

Page 3
Your original log sheet.

Page 1 should include:

- A statement of the hypothesis being tested.

- A description of the experiment you designed to test the hypothesis.

- A description of where, when, and how you made the measurements, including uncertainties.

- Your conclusions, especially with reference to the hypothesis. Also, compare measurements for the various light sources you tested.

Log Sheet: Solar Power

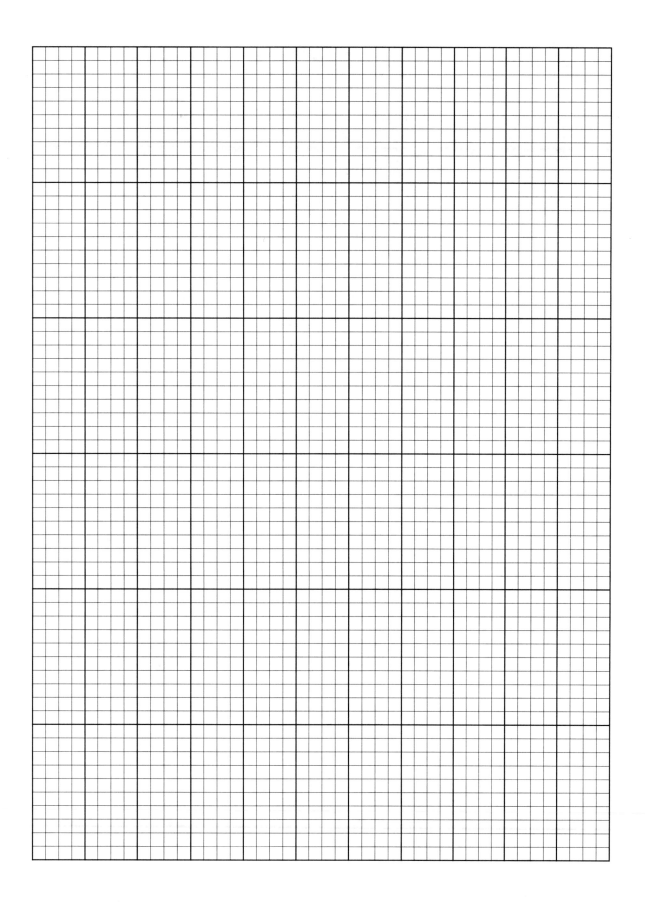

Activity 7
Air Pollution
Acid Deposition

What Is Acid Deposition?

When liquid water emerges from the air (from water vapour), it may appear as dew or fog, or it may fall to the ground as rain, snow, or hail. Dew and fog are forms of **condensation**, while rain, snow, and hail are forms of **precipitation**. The purity of condensing or precipitating water, regardless of its form, depends upon the quality of the air from which it came. Normal air contains carbon dioxide which, when mixed with water, leads to carbonic acid. Thus, normal condensation or precipitation is somewhat acidic. However, the burning of fossil fuels injects pollutants into the air which may significantly enhance the acidity of the water. For example, sulphur dioxide gas goes up into the atmosphere when coal is burned (say, to generate electricity). When mixed with water in the air, sulphuric acid results. Cars which burn gasoline expel nitrous and nitric oxide gases. When mixed with water in the air, nitric acid forms. Ultimately, condensation and precipitation deposit acidified water on the surface of the Earth. Even dry forms of acid may fall out of the air. All of these processes are referred to as **acid deposition**. Acidic precipitation in the form of liquid water (as against ice) is known as **acid rain**.

Why Does Acid Deposition Matter?

Some degree of acidity in precipitation is normal, so one might ask why more acidity is bad. The problem is that living things have adapted to modestly acidic precipitation. When the chemical composition of the precipitation deviates significantly from what is normal, environmental chemistry is modified. Acids are highly reactive chemicals. When mixed with the environment, they can have a deleterious impact upon living things. For example, acid deposition can impair photosynthesis, the process by which plants extract energy from the Sun for their survival. Deposition onto soil leads to the leaching of nutrients, thereby inhibiting the productivity of plants dependent upon those nutrients for their health. In turn, this can lead to reduced resistance to cold, disease, insects, and drought. Runoff concentrates acidified precipitation into lakes and rivers, thereby altering the chemistry of aquatic ecosystems. Lakes can become incapable of supporting life. Acid deposition is also known to damage the façades of buildings. It can even leach metal from pipes, contaminating the water supply for humans. Finally, acidified water vapour and acid-coated particulates are major components of smog, which is known to cause respiratory illnesses.

What Is Acidity?

Water's acidity is related to the preponderance of ions mixed up with water molecules. **Ions** are atoms or molecules which are charged as a result of having lost or gained one or more electrons. The two most important ions insofar as acid deposition is concerned are the hydronium ion (H_3O^+) and the hydroxyl ion (OH^-). At any instant, liquid water is composed not only of regular water molecules, but also of hydronium ions and hydroxyl ions which are produced when water molecules split apart (naturally). In pure water, there are equal numbers of hydronium and hydroxyl ions. Such a mixture is said to be **neutral**.

Certain chemicals, when mixed with water, break up into pieces which lead to an enhancement of the number of hydronium ions. The number of hydronium ions ends up exceeding the number of hydroxyl ions, and the mixture is said to be **acidic**. The chemical causing the effect is called an **acid**. Vinegar is an example.

The opposite may also occur. Other chemicals, when mixed with water, break up into pieces which lead to an enhancement of the number of hydroxyl ions. The number of hydroxyl ions ends up exceeding the number of hydronium ions, and the mixture is said to be **alkaline** or **basic**. The chemical causing the effect is called a **base**. Baking soda is an example.

Whatever way the imbalance goes, a solution with unequal numbers of hydronium and hydroxyl ions is highly reactive. The danger of acids has become well known. Perhaps less well known is that bases are among the most powerful and hazardous of household chemicals (witness drain cleaners).

How Do You Quantify Acidity?

To quantify the reactivity of a mixture, scientists measure the number of hydronium ions per litre and compare it with what is expected for normal water. On this basis, the mixture is assigned a rating, called the *pouvoir hydrogène*, or **pH**. A neutral mixture, which is one where there are just as many hydronium ions as hydroxyl ions, is assigned a pH of 7. An acidic mixture, which is one where there are more hydronium ions than hydroxyl ions, is assigned a pH less than 7. A basic mixture, which is one where there are fewer hydronium ions than hydroxyl ions, is assigned a pH more than 7. Note that every step of the scale corresponds to a change by a factor of 10 in the number of hydronium ions per litre. Thus, a mixture with a pH of 5 has 100 times more hydronium ions per litre than a mixture with a pH of 7.

Normal precipitation has a pH around 5.6, which is somewhat acidic. The term "acid deposition" is reserved for precipitation which has a pH **lower** than 5.6. Fish can't survive in water with a pH lower than 4.5. For reference, the pH inside your stomach is between 1 and 3, and the pH of oven cleaner is around 13. Fog in California has been reported to have a pH as low as 2! The pH of drinking water is close to 7 (neutral), how close depending upon the source and the chemicals used to treat it.

Scientists have developed dyes which change colour when exposed to acids or bases. To make the measurement of pH easy, these dyes are impregnated into **pH test strips**, which are usually made of paper or plastic. To measure the pH of a mixture, such as rain water, all you have to do is wet a pH strip with the mixture. The mixture will cause the strip to change colour in a way dependent upon the number of hydronium ions per litre. The colour defines the pH.

What You Should Do

Design and execute an experiment to test the following hypothesis:

My immediate environment is being subjected to acid deposition.

To begin, put a clean and dry container outside for as long as possible in order to catch some rain or snow. If none comes soon enough, clean the container again and go and collect some water from a *natural* body, such as a lake, pond, or puddle (not what comes out of a hose, for example, which is tap water). Whatever you do, keep notes as you go along using the included log sheet.

Using up to 6 of the pH strips included with your kit, measure the pH of the water you collect and the pH of the water coming out of your tap. Also measure the pH of at least one mixture of your own choice. Juice from a fruit or vegetable? Your own saliva? Whatever interests you? All records should be entered in pen on the included log sheet.

What You Will Need

From Kit

- Up to 6 pH test strips (you should leave 4 for other activities).
- Clean plastic test tubes to hold rain water or melted snow for testing (you decide how many).

From Other Sources

- At least one clean container for collecting rain or falling snow, preferably a bucket.

Things You Should Think About

- A truly useful experiment is one which is controlled, statistically significant, and repeatable. Make a logical choice as to what to compare your pH measurements against.

- Depending upon how you clean containers, contaminants may be introduced into the experiment which affect the pH of your samples. Make sure your containers are completely dry before you use them. So as not to introduce new variables, you should develop a uniform procedure for cleaning.

- Once it hits the ground, the chemistry of precipitated water may change. For example, puddles at the side of the road may be contaminated with pollutants accumulated during runoff, such as salt laid down to melt snow. If you need to sample a natural body of water, it is best to go off-road.

Activity 7   Air Pollution

What You Should Disseminate

Page 1
Describe your experiments and your findings.

Page 2
Show a detailed drawing of your experimental setup for measuring the pH of precipitated water. It should be carefully and liberally labeled.

Page 3
Your original log sheet.

Page 1 should include:

- A statement of the hypothesis being tested.

- A description of the experiment you designed to test the hypothesis.

- A description of where, when, and how you made the measurements, including uncertainties.

- Your conclusions, especially with reference to the hypothesis. Also, compare the pHs of the various liquids you have tested.

How to Use the pH Test Strips

Your kit includes 10 pH test strips, up to 6 of which you are allowed to use in this activity. In the middle of each strip is a yellowish rectangle whose colour is sensitive to pH. Numbered rectangles on either side show the correspondence between colour and pH. One way to measure pH is to wet only the pH-sensitive rectangle, say using your pipette. However, the perception of colour may be affected by the liquid especially if the liquid isn't clear. A better approach, where possible, is to immerse the entire strip in liquid, say by inserting it into a test tube full of liquid. This will guarantee that the colours of the pH-sensitive area and the pH scale are perceived in the same way.

To carry out a measurement of pH, you should first ensure you have clean dry hands. Get a pH strip and hold it by one end. Take care not to let your fingers spend too long on the part of the strip you will be exposing to liquid, as sweat can mess it up. Wet the strip with your liquid and wait ten seconds, that is until the colour of the pH-sensitive area has stabilized. Then compare the colour of the pH-sensitive area with the colour chart to judge the pH.

Log Sheet: Acid Deposition

Activity 8
Intensive Agriculture
Soil Quality

What Is Intensive Agriculture?

The phrase **intensive agriculture** refers to modern farming practices which make use of large inputs of fertilizer, water for irrigation, and fossil fuels. Combined with seeds selected or engineered for rapid growth and high grain yield, these techniques have resulted in a remarkable increase in the amount of food available to each person over the past fifty years. However, this increase has come at the price of significantly increased reliance on artificial fertilizer to combat ever decreasing soil fertility.

A fundamental problem with farming of any sort is that, when a crop is harvested, the nutrients that have been taken up from the soil by the crop plants are carried away from the field. Some of these nutrients then end up being consumed by humans, but this leaves fewer nutrients back in the field, weakening its fertility. Thus, nutrients must be replenished using fertilizer.

Traditional organic fertilizers, such as animal manure, have the virtue that they help to maintain the structure of soil, but they are also difficult to control precisely. High-yielding seeds respond best to specific amounts of fertilizer delivered at just the right time during the growing cycle. Artificial (chemical) fertilizers can be applied in precise amounts as needed. Unfortunately, long-term use of artificial fertilizers with little input of organic matter tends to degrade the soil, making it drier and more acidic. In fact, its utility becomes completely dependent on injections of fertilizer.

What Is Soil, and What Determines Its Quality?

Soil is one of the most important components of any terrestrial ecosystem. It is the medium that supports and nourishes the Earth's **producers**, that is the blanket of vegetation that stores energy from the Sun in the form of carbohydrates (via photosynthesis). Soil is, therefore, a crucial underpinning of the food supply of all animal life.

Soil is actually a mix of mineral and organic constituents. The mineral constituents, essentially little bits of eroded rock, can be distinguished by the size of the bits: the biggest particles are called **sand**, the smallest are called **clay**, and the range in between includes mid-sized particles of **silt**.

Different types of soil are characterized by the relative proportions of these three components.

However, not all soils are equally good for supporting plant life. A good soil has just the right proportions of clay, silt, and sand to allow enough water to be held between the small particles to supply plant roots, and to allow for enough air between the larger particles so that those same roots can breathe. Generally an even mix of sand, silt, and clay, called **loam**, is best.

Different soils are also distinguished by the amounts of **inorganic nutrients** they contain. These nutrients are naturally cycled through the environment. They are taken out of the soil as plants grow, but they are returned to the soil when plants or animals die and their organic matter decays. For plants, the most important inorganic nutrients are **nitrogen** (in the form of nitrate ions, NO_3^-), **phosphorus** (as phosphate ions, PO_4^{3-}), and **potassium** ions, K^+ (recall that an ion is any atom or molecule that carries a net electrical charge). These nutrients are taken up from soil water through the roots.

The amounts of inorganic nutrients in a given soil depend on the amount and rate of decay of the organic matter and on the rate at which plants take in the nutrients. For this reason, nutrient concentrations vary with time during the year and during the growing cycle.

Once released into the soil, negative ions like phosphate and nitrate are very susceptible to being washed out of the soil when water passes through. This process is called nutrient **leaching**. Thus, it is preferable that the supply of nutrients in the soil not exceed the plants' demand by too much. Otherwise, rainfall is likely to leach away nutrients before they can be used. Natural soils accomplish this balance because the release of nutrients from decaying organic matter is slow.

Why Should You Care about Soil Quality?

Declining soil quality means fewer and/or smaller crop plants. In either case, the end result is a decrease in the amount of food available to each person. In other words, soil degradation poses a long-term threat to our food supply.

In the short term, to combat declining fertility, intensive farming uses large inputs of chemical fertilizer to provide the major nutrients. However, excessive and/or inappropriate use of fertilizer can result in large quantities of excess nutrients being leached out of farm fields. Eventually the nutrients get into waterways where they become pollutants.

Finally, but on a more cheerful note, an improved understanding of soil may also make you a better home gardener.

What You Should Do

Of all the activities in this kit this one is the most involved. It has three major parts. Before you do anything else **READ ALL OF THE INSTRUCTIONS. You should plan to perform the different steps in this activity over *two* consecutive days.**

Obtain two soil samples (as explained below in *What You Will Need*) and label the one that is most topsoil-like *A* and the other *B*. Next, divide sample A into two roughly equal parts. Make sure that both soil samples are dry before you begin any testing.

Soil Sedimentation

Examine the physical structure of one half of sample A. To do this, put the sample into a container that can be securely sealed (a glass jar with a cover for example, or a water bottle). **It is crucial that the sides of the container be transparent.** Fill the container with at least twice as much water as there is soil (use less soil if necessary to accommodate water). Seal the container and shake it *vigorously* for at least one minute. Then set the container aside and let the solids settle for at least one hour. Big sand particles settle immediately, but very fine and much lighter clay particles can remain suspended in the water for much longer.

Once the soil has settled, there should be only bits of undecomposed organic debris floating at the top of the water. **DO NOT SHAKE THE CONTAINER ANY FURTHER!!** Looking through the sides of the container, examine the soil that has settled and record what you see. You may need to shine a bright flashlight onto the jar —bright sunlight also works well. Look for any visible layering and record the visual characteristics of the different layers (relative size of grains, colour, etc.). Using a ruler, measure each layer's thickness.

Water Holding Capacity

Using the second half of sample A, measure how much water can be held by a sample of topsoil. To do this, you have to wet dry soil with a known amount of water, say W_1, and measure the amount of water that **isn't** absorbed by the soil, W_2. The amount of water held by the soil, i.e., the **holding capacity** C, is the ratio of what is held to what went in . Thus, $C = (W_1 - W_2)/W_1$.

The kit includes a few coffee filters to help you perform the measurements, but it is up to you to design the experiment. Use a volume of water which is a few times greater than that of your soil sample. If you use too much, you will not be able to measure C very well, nor will you be able to detect any effect on the runoff in subsequent experiments. Note that you can quantify the amount of water before and after it has passed through the soil by measuring either its volume (with a measuring cup) or the height the water reaches in a fixed container (with a ruler).

Ideally you should **wait 24 hours after you have wet the soil before measuring the water that came out**. Cover the soil sample and store it in a cool dark place during this time to minimize evaporation.

Water that has passed through the soil is called the **leachate**. It will be the subject of the next tests, so **DO NOT DISCARD THE LEACHATE**.

Using a second container, obtain the leachate from a similar quantity of soil sample B. This time you don't need to make any holding capacity measurements (but you can if you want). You will need this leachate in the next step.

Acidity, Nutrients, and Leaching

Using both of the leachates obtained previously, test the following hypothesis:

Passage through soil can change the pH of water.

In measuring the pH of the leachate, you are in fact measuring the pH of the soil. Refer to the *Air Pollution* activity if you don't remember how to use the pH strips.

The implication of the above hypothesis is that water can leach nutrients out of soil. If this is true, then the following hypothesis is motivated:

Water leaches nutrients from soil.

Test this hypothesis by performing two controlled experiments, one for nitrate and one for phosphate, using the test tablets and test tubes provided in your kit. Test both of your leachates to see if there is any difference in nutrient content between the two soils. Detailed instructions on the use of the nitrate and phosphate test tablets are provided below.

What You Will Need

From Kit

- Sandwich bags (for collecting soil)
- Coffee filters
- Elastic bands to secure the coffee filters to the containers
- Plastic spoon
- pH paper (up to 4 strips)
- Plastic pipette (looks like an eye-dropper)
- Test tubes with plug caps
- Nitrate test tablets
- Phosphorus test tablets
- Ruler

From Other Sources

- Two samples of dry soil taken from two different locations. Enough to fill a sandwich bag with each sample should be sufficient. At least one sample should be of topsoil or something similar (this sample should be the one labeled A). The other sample can be potting soil, taken from a house plant if necessary. If your samples aren't dry, then spread them out on newspapers for a few days until they are dry. Do not dry soil in an oven as this may change its characteristics. Also, pick out any large stones and organic debris such as roots and leaves.
- Containers or jars with openings wide enough to accommodate the coffee filters (you decide how many).
- Lids for the containers
- Measuring cup (optionally)

Things You Should Think About

- Keep a record of where you obtained your soil samples. Was it your mother's garden, the side of the road, or store-bought? Was anything growing in this location?
- If doing this activity in winter, then avoid taking soil from the side of the road where salt concentrations may be high.
- For water-holding capacity, carefully measure and record the amount of water that you pour through the soil. A good amount here would be roughly twice as much water as there is soil. The amount should be sufficient to allow you to make good measurements using either a measuring cup or a ruler.
- When pouring water through the soil, do it slowly and mix the soil gently to make sure that it gets thoroughly wet. Don't leave any dry parts. Mix the soil a few times as you pour the water.
- If you choose to measure the amount of water using a ruler, be sure to use only one size of container to hold the water. The height of a given volume of water in a wide container will be different from the height in a narrow one.

What You Should Disseminate

Page 1
Describe your explorations, experiments, and findings regarding sedimentation and water-holding capacity.

Page 2
Describe your experiments and findings about soil acidity and nutrient content.

Page 3
Show a detailed drawing of your setup for performing the water-holding capacity experiment.

Page 4
Your original log sheet.

Page 1 should include:

- A summary of your investigation of the structure of soil sample A. In particular, sketch a picture which shows the different layers after the soil has settled in the container. Also, indicate the thickness of the layers.

- Your answers to the following questions: What fraction of your sample is made up of sand? How much is silt? How much is clay?

- A summary of your findings with regard to the water-holding capacity of soil sample A.

Page 2 should include (for each test):

- A statement of the hypothesis being tested.

- A brief description of the experiment you designed to test the hypothesis.

- A description of where, when, and how you made the measurements.

- Your conclusions, especially with reference to the hypotheses. How does the pH of the two soil samples differ? How do their nutrient contents compare?

How to Use the Test Tablets

Your kit contains tablets for testing the nitrate and phosphate content of a liquid. It is important that these be used correctly and safely to ensure accurate results. The chemicals aren't especially dangerous, but they are hazardous if ingested. **DO NOT EAT THE TABLETS NOR INGEST ANY LIQUID THAT HAS BEEN IN CONTACT WITH THE TABLETS. IF YOU DO, THEN CONTACT A DOCTOR IMMEDIATELY!** Among other things, this means that you shouldn't conduct experiments at the same time you are eating lunch.

To test the nitrate or phosphate content of a liquid, follow this procedure (note that you need a separate sample of liquid for each test!):

1. Using the pipette, transfer some of the liquid to a test tube. For a nitrate test, fill the test tube leaving about a centimetre at the top. For a phosphate test, transfer *25 drops* of the liquid into the test tube and fill the rest with warm tap water, again leaving about a centimetre at the top.

2. For a nitrate test, open one foil packet labeled "NITRATE WR CTA" and drop the tablet into the test tube. Try

Activity 8   Intensive Agriculture

not to touch the tablet. For a phosphate test, do the same, but with one of the packets labeled "PHOS". (If you do touch a tablet, then avoid touching your eyes or mouth, and wash your hands with soap and water as soon as you can.)

3. Seal the test tube with one of the plug caps and shake until the tablet completely dissolves.

4. Allow five minutes for the colour to develop. In the presence of nitrate ions, the liquid will turn pink with a deeper shade of pink indicating a greater amount of nitrates. In the presence of phosphate ions, the colour will be blue, with a deeper blue indicating more phosphates. Don't wait too long to make your observation.

5. After you have made your observation, pour the liquid in the test tube down the drain. Rinse your test tube with warm clean water before performing another test. **Do not use soap. Otherwise, you may contaminate your test tube and mess up further tests!**

Log Sheet: Soil Quality

Activity 9
Genetically Modified Organisms
Genetics

What Are Genetically Modified Organisms?

A **species** can be regarded as a population of organisms which can reproduce freely in the wild. **Traits** are characteristics of a species which are inheritable through reproduction. For example, the colour of your eyes is a trait. Different expressions of traits are called **phenotypes**. For example, your eyes may be blue or brown. Phenotypes are determined at the microscopic level by chemicals called **proteins**. In living things, proteins are manufactured using instructions stored inside cells.

In the past, traits in offspring were determined solely by parents. This meant that traits in one species could never be transferred to another species, because interbreeding, if in fact it were possible, would lead to infertile offspring. For example, although it is possible for a horse to mate with a donkey, the offspring, which is a mule, cannot reproduce. Crossing a cat with a tree can't even be contemplated. Thus, a new variety of a species could only be promoted by tapping into the existing suite of traits found already in the species. For example, if you wanted to create a herd of blue pigs, you would first have to find a blue pig.

Today, by focusing on the molecular basis of life, biotechnology is enabling crossovers. Through genetic engineering, it is in principle possible to transfer a trait of any living thing to another regardless of the species. For example, it is possible to create a glowing monkey by inserting a jellyfish's code for bio-luminescence into an egg cell of a monkey. A living thing with a bio-engineered trait is referred to as a "genetically modified organism," or **GMO** for short.

What Is the Problem?

Genetic engineering is revolutionizing society in ways which could never have been predicted. As an example, it is now possible to create bacteria which can eat up oil spills. Farmers can grow genetically modified corn which is resistant to herbicides. Experiments are underway to produce a banana which delivers a vaccine for Hepatitis B and a sunflower which produces useful amounts of rubber.

So, why are GMOs an environmental issue? The reasons are many. First, new proteins are being introduced into the food supply which have never been there before. Allergic reactions do occur, but appear to be rare. However, it is unclear what the long-term consequences to human health may be. There is some concern that herbicide-resistant plants may spread their resistance to weeds. If so, a generation of super-weeds might take over the landscape. Some people worry that the creation of "foodaceuticals" like the bio-engineered banana might lead to more virulent viruses as existing viruses adapt to the new environment. There are also ethical issues still to be confronted. If it were possible to bio-engineer a cow to produce milk like that from a human mother, would it be right to eat the beef?

Fundamentals of Genetics

From the above discussion, it should be apparent that GMOs are a significant environmental issue. Thus, it is important to understand their origins. To do so requires some fundamental knowledge about genetics.

The structural unit of living things is the cell. In the nucleus of every cell are **chromosomes**. Each chromosome is one giant molecule called **DNA**. DNA looks like a spiral staircase. Different segments of the DNA molecule, i.e., different sets of stairs, give instructions on how to make a protein for a particular trait. One such segment is called a **gene**. In essence, then, the nucleus of a cell is a library of libraries. Each chromosome (DNA molecule) is a library of instruction manuals, with each manual being a gene. All cells in an organism carry identical copies of the chromosomes.

To create an organism with a new trait, the gene which contains the instructions for producing the relevant protein must be identified in the DNA of the species already exhibiting the trait. The gene is extracted and inserted into the DNA of a bacterium. Bacteria multiply rapidly, so it doesn't take long before an entire culture hosts the new gene. The genes are harvested and injected into a culture of cells from the organism to which the trait is to be transmitted. After becoming integrated into chromosomes, cell division ultimately leads to an organism with the new trait.

Cells in human beings contain 46 chromosomes arranged in 23 pairs. Of every pair, one chromosome is from your father and one is from your mother. In other words, any particular chromosome from your father has a counterpart from your mother. However, the counterpart from your mother is not identical to that from your father; there may be subtle differences between genes. Different forms of a gene for a particular trait are called **alleles**. For example, your father may have contributed a gene which causes freckles, but the corresponding gene from your mother may stop them. Whether or not you end up with freckles (your phenotype) depends upon which allele is more powerful with respect to production of the protein responsible for freckles, i.e., which is **dominant** versus which is **recessive**.

Tracing Genes

In the case of freckles, the gene which leads to freckles is dominant. This means that if neither of your parents has the gene for freckles, then no offspring can be freckled. Label the gene for freckles F and the gene for no freckles f. What if your father had one F gene and one f gene, but your mother had two f's? You get only one gene from each parent. It is possible that you could end up with F and f or f and f. If F and f, then you would have freckles, because F wins out over f. However, if you ended up with f and f, you wouldn't, because the protein for freckles can't be produced.

The pair of genes which determines how a particular trait is expressed defines a **genotype**. In the case of freckles, there are four possible genotypes: FF, Ff, fF, or ff. Here, the first letter refers to the allele from your father and the second to the allele from your mother. Since F is dominant, then freckles will be realized in 3 of the 4 cases.

If a trait is expressed in different ways among your siblings, e.g., some have freckles and some don't, then at least one parent carries the F gene and both parents carry the f gene (because you need one f from each parent to get no freckles). In other words, at least one parent has to be a hybrid, namely Ff or fF.

An example of a genetic tree for freckles is shown in Figure 5. Squares are used to mark males and circles are used to mark females. Phenotypes are identified for each person. Allowed genotypes are written underneath, based upon analysis of the lineage.

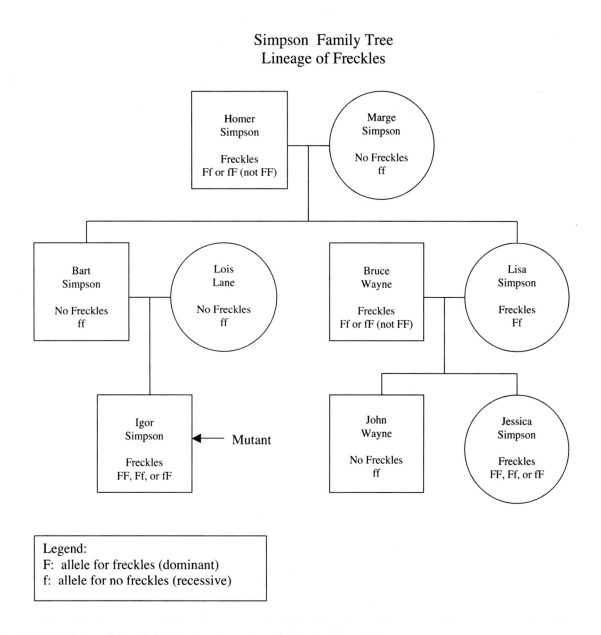

Figure 5: Example of a genetic family tree.

What You Should Do

Trace the phenotypes and genotypes of a trait in your family. The best kind of trait to trace is one which has only two possible expressions. This is because the trait's genetic heritage is simple, in that there is a good chance that only one gene is behind it. More complicated characteristics like eye colour, which has many expressions, are controlled by multiple genes, so it is very difficult to judge genotypes.

Here are some suggestions as to what to trace.

- Ear lobes. They can be attached or free. A free ear lobe is one which has a flap of skin hanging down from the ear which is not connected to the face. The allele for free ear lobes is dominant.

- Widow's peak. This is a V-shaped extension of hair toward the nose at the top middle of the forehead. The allele for having it is dominant.

- Clasped hands. Which thumb ends up on top is determined by genetics. The allele for left on top is dominant.

- Tongue curling. This is an ability to curl the edges of the tongue upwards. Either you can do it or you can't. The allele which gives you the ability to do it is dominant.

- Freckles. As described above, the allele for having freckles is dominant.

- Dimples. Either you have them or you don't. The allele for having dimples is dominant.

- Hitchhiker's thumb. This refers to an ability to bend the thumb backwards and outwards (toward the nail-side of the thumb). Some people can only extend their thumb straight out. The allele which leads to hitchhiker's thumb is recessive.

- Bent little fingers. If you put your hands together side by side with palms up, the tops of the little fingers will either be parallel or bend away from each other. The allele for bending away is dominant.

From a genetic standpoint, it is most interesting to trace a trait for which both expressions are present in your family. For example, if nobody in your family has freckles, there is really nothing to be learned by tracing the trait.

After picking a trait to trace, contact members of your family and gather information about phenotypes as far as you are able. Include as many relatives (branches) as possible. Records should be entered in pen on the included log sheet. Then, draw a family tree, labeling each branch with the name of the individual and how the trait is expressed (e.g., "freckles", or "no freckles").

Knowing which allele for the trait is dominant, do your best to assign a genotype to each family member, and write it down on the tree. Use a capital letter to describe the dominant allele, and a small letter to describe the recessive allele. If there is more than one possibility for a genotype, give all possibilities.

Finally, imagine you were to have a child with someone who displays the recessive version of the trait you have traced. With a dotted line, draw the child on the family tree. Write down your predictions for the child's genotype and phenotype.

What You Will Need

From Other Sources

- Phone

- Family photographs

Things You Should Think About

Occasionally a phenotype is inconsistent with a genotype. For example, suppose that an offspring of two parents without freckles ended up with freckles. This would appear to be impossible, as neither parent hosted the allele for freckles. So, how could it happen? It could be that there is another gene which in rare circumstances has a bearing on whether or not someone has freckles. Alternatively, it is possible that during the development of the embryo there was a mutation, i.e., a spontaneous change in the DNA, although it would be highly unlikely that such a mutation would just happen to lead to freckles. Another possibility is that there is something unusual about the relationship between the offspring and the parents. Or, perhaps the environment plays a role, say through what a child is fed. If you encounter this situation, be sure to make note of it and explore possible explanations.

What You Should Disseminate

Page 1

Your family tree, as far back as you can easily trace phenotypes (for example, grandparents. First, it should identify the trait being traced and which allele is dominant. It should distinguish between males and females with squares and circles. It should identify the name and phenotype of each family member. Also, it should display possible genotypes for each family member, with a legend describing your choice of lettering (what capital and small letters refer to). Finally, it should show the predicted genotype and phenotype of any child you have with an individual who displays the recessive version of the trait.

Page 2

Your log sheet.

Log Sheet: Genetics

Activity 10
Radiation
Biological Effects of Radiation Exposure

What Is Radiation?

To many people, radiation is an invisible demon. More often than not, it brings forth visions of nuclear disasters and the health implications stemming from them. What is generally not appreciated is that radiation is a part of life, and may even have had a role to play in the evolution of life. Whether deciding to have an X-ray or judging the merits of nuclear power, it is important to have a basic understanding of what radiation is and what it can do to be able to appreciate the risks or benefits of exposure.

The word *radiation* is used in many contexts, so in itself can be confusing. What most people think about when they hear the word radiation is something which is hazardous to life. Light is sometimes referred to as radiation, which is a short way of saying *electromagnetic radiation*. However, few people regard the emanations of a reading light as being dangerous. Rather, radiation to worry about consists of **light waves or particles of matter with energy sufficient to strip atoms or molecules of surrounding electrons**. Such radiation is more explicitly denoted as **ionizing radiation**, because the stripping process is known as **ionization**. Only the most energetic forms of light, X-rays and gamma rays especially, qualify as ionizing radiation. Fast-moving particles of matter also count as ionizing radiation, the most common of which being helium nuclei, referred to as **alpha particles**, and electrons, referred to as **beta particles**. In this activity, the word *radiation* is used as a label for *ionizing radiation*.

On Earth, radiation usually comes from the spontaneous disintegration of unstable atomic nuclei, i.e., atomic nuclei which seek out a lower state of energy. Atomic nuclei which can disintegrate spontaneously are said to be **radioactive**, and the actual act of breaking apart is referred to as **radioactive decay**.

An **element**, such as oxygen, iron, or uranium, is a building block of nature. The smallest piece of an element is called an **atom**. The behaviour of an element in nature, and thus the name of the element, is set by the character of its atoms – all have an identical number of protons in their nuclei. For example, an element whose atoms contain 27 protons each is called *cobalt*. Every element has more than one version, though, distinguished by the number of neutrons mixed up with the protons in each of its atoms. Each version is referred to as an **isotope** of the element. Isotopes which are unstable, and thus radioactive, are often referred to as **radio-isotopes**.

A common way of naming an isotope is to refer to the name of the element and the total number of protons and neutrons in each of its atomic nuclei. Thus, the most common isotope of cobalt is called *cobalt-59*. Nuclei of this particular isotope contain 32 neutrons along with the 27 protons which define the element to be cobalt (the sum being 59).

Cobalt-59 is the only stable isotope of cobalt. A particularly interesting radio-isotope is cobalt-60, atoms of which have one more neutron in their nuclei. After a time, it decays by transforming one of its neutrons into a proton and an electron. A beta particle (the electron) moving at nearly the speed of light and two gamma rays are released, leaving behind a stable nucleus of nickel.

What makes cobalt-60 interesting is the time scale over which it decays. In general terms, the rate at which a radio-isotope decays is characterized by how long it takes for half of the nuclei in a sample to disintegrate. This is referred to as the **half-life** of the nucleus. The slower the rate of decay, the longer is the half-life. Cobalt-60 has a half-life of 5.3 years, which is short enough (meaning that the decay rate is fast enough) that relatively small samples can be used to provide a useful supply of radiation for a wide range of applications. For example, cobalt-60 is used to detect flaws in metal pipes, to kill bacteria in food, and to treat various kinds of cancer. On the other hand, cobalt-60 is an important constituent of nuclear waste, being produced when radiation inside a nuclear reactor interacts with the steel walls of reactor components such as fuel rods. The half-life is great enough that radiation from cobalt-60 continues to be a concern long after a fuel rod is spent.

How Is Radiation Quantified?

Radiation carries energy, and the influence of radiation striking matter is rooted in how much energy is transmitted to the matter. The amount of radiation energy impinging upon a body of matter is referred to as the **exposure**. The amount of energy actually absorbed by the body is called the **absorbed dose**. The absorbed dose is quantified by the **gray:** 1 gray (Gy for short) = 1 Joule per kilogram, where the Joule is a measure of energy. In other words, if each kilogram of a tree absorbs 1 Joule of radiation energy, then the tree has received an absorbed dose of 1 gray. An older description of the absorbed dose is the **rad**: 1 rad = 0.01 Joule per kilogram. Thus, 1 gray is equivalent to 100 rads.

The effect of radiation on human tissue depends not only on the amount of energy absorbed but also upon the type of radiation. For example, 1 gray from alpha particles is much more damaging than 1 gray from beta particles even though the same amount of energy is absorbed. To understand this, it is useful to regard the body as a fortress under bombardment. An alpha particle, which is a conglomeration of two protons and two neutrons, is like a cannon ball, and a beta particle, which is just an electron, is like a bullet. Cannon balls tend to cause a lot more damage than bullets. How damaging a given absorbed dose of radiation is to living tissues is called the **dose equivalent**. It is the absorbed dose multiplied by a *quality factor* which takes into account the constituents of the radiation. The dose equivalent is quantified by the **sievert**: 1 sievert (Sv for short) = quality factor × 1 gray. The quality factor for radiation composed of beta particles or gamma rays is 1, whereas the quality factor for radiation composed of alpha particles is 20. Thus, a human subjected to an absorbed dose of 1 gray of beta particles would be receiving a dose equivalent of 1 sievert, but a human subjected to the same absorbed dose of alpha particles would be receiving a dose equivalent of 20 sieverts. In other words, 20 times more beta particles would have to be absorbed to incur the same amount of damage as the alpha particles. An older way of quantifying the dose equivalent is with the **rem**: 1 rem = quality factor × 1 rad. Thus, for a given quality factor, 1 sievert is equivalent to 100 rems.

How Does Radiation Affect Living Things?

Although radiation damages living things at a microscopic level, the consequences can end up being manifested visibly. When a complex organic molecule is ionized, the way in which it interacts chemically with other molecules can change. As a result, the molecule's ability to participate in life functions may be impaired, if not defeated altogether. Admittedly, an organism houses a great many molecules, so damage to one of them would have no noticeable effect. However, if enough molecules are damaged, which can happen with intense or prolonged exposure to radiation, then the health of the organism or its offspring can be affected.

What matters most to the integrity of an organism is the health of its cells. Cells can be killed when exposed to radiation. If these cells make up bone marrow, for example, then immunity to disease will be reduced because fewer white blood cells will be produced. Of particular concern are the consequences of altering the DNA housed in cells. Exposure of DNA to radiation can lead to changes in its structure, some of which can alter the genetic code. A permanent modification to DNA is called a **mutation**.

Somatic mutations are uninheritable changes to DNA which lead to observable manifestations in the organism itself. One such manifestation is cancer, which arises when cell multiplication becomes uncontrolled. **Germline mutations** are inheritable mutations which arise when the DNA in a sex cell (spermatozoon or ovum, for example) is altered. Through cell division, such a mutation propagates from a fertilized egg to all cells of the offspring, and is manifested as a new trait (in the offspring). Of course, it is traits which underlie *survival of the fittest* by influencing an organism's ability to thrive or wither in any particular environment. If the trait promotes survival, reproduction will ultimately make it a characteristic of the species. By causing mutations, it is very possible that natural sources of radiation played a significant role in the evolution of species on Earth.

Where Is Radiation Coming From?

Radiation is everywhere. Even human beings are radioactive. Not because of technology, but simply because radio-isotopes are a part of nature. They are found in trace quantities everywhere, including your food and water and even the air you breathe. Also, you are constantly being bombarded by radiation from space, called **cosmic rays**, such as from the Sun and exploding stars.

What might be regarded as a safe dose of radiation for humans would be what you are exposed to from natural sources, which of course has been unavoidable since life's earliest beginnings. Something like 130 cosmic rays pass through you every second, delivering an absorbed dose of about 0.3 millisieverts per year at sea level (a millisievert is one thousandth of a sievert). The dose is greater at higher altitudes. Another 0.3 to 1.2 millisieverts per year come from naturally-occurring radioactive constituents of rocks and soil, which can end up in building materials, food, and water. One of them, uranium-238, decays to produce radon-222, which is a radioactive gas which seeps into our homes. Humans themselves contain trace amounts of about 400 radio-isotopes, such as potassium-40 and carbon-14. Every second, about 8000 nuclei decay in your body. Thus, you are self-exposed to about

0.4 millisieverts per year. Overall, from natural sources, you are receiving an absorbed dose of 1 to 3 millisieverts per year.

What is generally not appreciated is that human exposure to radiation generated by technology is normally substantially less than that from natural sources. From medical and dental X-rays, a typical dose would be 0.4 millisieverts per year (one set of dental X-rays contributes about 0.1 millisieverts). Another 0.03 millisieverts per year comes from the burning of fossil fuels, such as coal, oil, and gas. Nuclear weapons testing has left a remnant of 0.04 millisieverts per year. Radiation associated with normally-operating nuclear power plants typically delivers only 0.0002 millisieverts per year, although this could be as high as 0.01 millisieverts per year if you live next door to the plant. Overall, the amount of radiation received from technology could be around 1 millisievert per year, the majority of which being tied to dentistry and medicine. Thus, you can expect to be receiving a total radiation dose of from 2 to 4 millisieverts per year, with 50% to 75% of it coming from natural sources. A dose of 3 millisieverts, which is the worldwide annual average, is equivalent to about 30 dental X-rays.

What is dangerous? A dose of a million millisieverts (100,000 rems) leads to immediate death. A whole-body dose of 5000 millisieverts (500 rems) will kill 50% of people exposed to it (in a couple of weeks). However, the same sort of dose targeted at a cancerous organ can save a life. A dose of 1000 milliseiverts (100 rems) can cause vomiting, fatigue, abortion of early pregnancies, and temporary sterility in males. Also, there is observed to be an increased probability of leukemia. Early embryos can display abnormalities after being exposed to 100 millisieverts (10 rems). Below about 10 millisieverts (1 rem), though, it has proven difficult to detect any deleterious effect on health.

Mutations to DNA are cumulative, so in principle even exposure to low levels of radiation could be dangerous if the time of exposure is long enough. However, the effects of long-term exposure to weak sources of radiation are very difficult to observe, because there are so many variables which affect the health of a human over a lifetime. What is known is that there are biological processes which can repair damage to DNA, so not all changes induced by radiation end up as mutations (i.e., permanent). Whether or not a change to DNA becomes permanent depends upon the rate at which the damage occurs relative to the capacity of the body to repair itself. This means that a given equivalent dose received slowly (say, over a lifetime), could be less hazardous than the same dose received quickly (say, in a day), because more incidents of DNA damage would be likely to be repaired.

How Might You Explore the Effects of Radiation?

Radio-isotopes tend to be dangerous, and are not items which can be readily included in a kit such as this. Thus, it is not possible for you to play around with radiation itself. However, it is possible to get a hold of organic materials which have been exposed to radiation. They aren't radioactive, but they can be studied to evaluate how radiation may have affected them.

In your kit, you will find two sets of radish seeds. One set of seeds has been exposed to no more radiation than typically emanates from the environment. The second set of seeds has additionally been exposed to beta and gamma rays from cobalt-60. The absorbed dose was 1,500 grays

(150,000 rads). For humans, the quality factor for beta and gamma rays is unity, so the equivalent dose would be 1.5 million millisieverts (150,000 rems). In other words, humans would face instant death if exposed to the amount of radiation to which the radish seeds in the second sample have been subjected.

Seeds are in many ways like eggs. Upon germination, cells divide and multiply, ultimately leading to plants made up of cells with the same genetic code as carried by the original seeds. If radiation damages the cells in the seeds, then effects could be manifested not only by the health of the plants but also by their traits.

What You Should Do

Design and execute a useful experiment to test the following hypothesis:

Seeds exposed to an excessive amount of radiation yield abnormal plants.

Obviously, you will want to attempt to grow some radish plants with the seeds provided. **Be aware that a period of at least 4 weeks will be required to grow your plants, and that you must plan your work accordingly.**

You will need soil, one or more containers with small holes in the bottom to hold the soil and drain excess water, and one or more platforms on which to rest the containers to collect the excess water (to protect your furniture). Whatever containers you use should be at least 10 cm deep to allow good root (and radish) growth. They should be wide enough to allow several seeds to be planted at least 2 cm apart (otherwise you have to use more containers). Plastic pots for plants can be acquired very cheaply at gardening centres or hardware stores. However, you can use something like margarine containers as substitutes if you cut some holes in the bottom of them.

Fill a container with damp soil. To plant a seed, make an indentation in the soil about 1 cm deep, insert the seed, and then cover it up. Plant as many seeds as your container will hold (given that the seeds have to be at least 2 cm apart). Then create a little greenhouse by attaching a piece of clear plastic wrap to the top of the pot with tape or an elastic band.

To grow radishes successfully, you will need lots of light, especially light with an ultraviolet component (which promotes photosynthesis). Sunlight or a fluorescent light will work (an incandescent light won't). Either place the radishes next to a window which lets in light from the sky (not necessarily direct sunlight, although that would be better), or place them under a desk lamp equipped with a fluorescent bulb. In the latter case, consider leaving the light on 24 hours a day.

Once seeds germinate and grow a bit, or if you see a lot of mould building up, remove the plastic wrap from the pot. Mould will go away once you expose the plants to air. You should make an effort to water your plants once every couple of days once the plastic wrap is removed.

After about 4 weeks, dig up whatever plants have grown. The roots may provide further insights into the validity of the hypothesis. When you are done observing, make a salad!

Evaluation of the hypothesis hinges entirely upon observation. First and foremost, you must ensure you label things well enough so that you know which plants grow from normal seeds and which plants grow from ir-

Activity 10   Radiation

radiated seeds. You should record regularly and carefully what you do and what you see from the beginning to the end of your experiment. If you feel it is useful or helpful towards evaluating the hypothesis, you should make drawings or take pictures. You might even consider measuring the plants in some way to track the progress of growth. All of your observations should be recorded on the supplied log sheets (use both sides).

What You Will Need

From Kit

- At least 10 radish seeds which *have not* been exposed to radiation from cobalt-60. These are provided in a *white* plastic bag.

- At least 10 radish seeds which *have* been exposed to radiation from cobalt-60. These are provided in a *coloured* plastic bag.

- Ruler

From Other Sources

- One or more containers and pedestals suitable for growing plants from your seeds (you decide how many)

- Soil. Potting soil will work.

- Plastic wrap

Things You Should Think About

- As usual, a truly useful experiment is one which is controlled, statistically significant, and repeatable. Especially, beware of other variables besides the exposure to radiation which might influence the outcome of your experiment. Whatever you do with the normal radishes should also be done with the irradiated ones (with the possible exception of eating).

What You Should Disseminate

Page 1
Describe your experiment and your findings.

Page 2
Show a detailed drawing of your experimental setup. It should be carefully and liberally labeled.

Page 3 and beyond
Your original log sheets and any support materials (such as drawings or photographs) which you think are vital to your investigation and its conclusions.

The first page should include:

- A statement of the hypothesis being tested.

- A description of the experiment you designed to test the hypothesis.

- A summary of your observations which are most relevant to evaluating the validity of the hypothesis.

- Your conclusions, especially with reference to the hypothesis.

Log Sheet 1: Biological Effects of Radiation Exposure

Log Sheet 2: Biological Effects of Radiation Exposure

Log Sheet 3: Biological Effects of Radiation Exposure